Hacker Cracker

Destined to Live, with William Ungar

Joycelyn Elders, M.D., with Joycelyn Elders

Never the Last Journey, with Felix Zandman

The Line of Fire, with William Crowe Jr.

Slow Burn, with Orrin DeForest

Into the Heart, with Kenneth Good

Warrior, with Ariel Sharon

In the Jaws of History, with Biu Diem

Portrait of the Enemy, with Doan Van Toai

The Vietnamese Gulag, with Doan Van Toai

A Vietcong Memoir, with Truong Nhu Tang and Doan Van Toai

Hacker Cracker

A Journey from the
Mean Streets of Brooklyn
to the Frontiers of Cyberspace

Ejovi Nuwere
WITH **David Chanoff**

WILLIAM MORROW

An Imprint of HarperCollins*Publishers*

Various names, places, and occasional details
have been changed or censored for reasons that
will be apparent to every attentive reader.

FIRST EDITION

Designed by Mia Risberg

Printed on acid-free paper

Library of Congress Cataloging-in-Publication Data
Nuwere, Ejovi.
    Hacker cracker: a journey from the mean streets of Brooklyn to the
frontiers of cyberspace / Ejovi Nuwere with David Chanoff.—1st ed.
    p. cm.
    ISBN 0-06-093581-2
    1. Nuwere, Ejovi. 2. Computer security—United States—Biography.
3. Computer hackers—United States—Biography. I. Chanoff, David.
II. Title.
QA76.2. N39 N89 2002
005.8'092—dc21
[B]

                                                    2002069611

05 06 WBC/RRD 10 9 8 7 6 5 4 3

TO MY MOTHER

Contents

Contents

Acknowledgments

I would like to thank my friends and family, especially my grandmother, for allowing me to publish our personal family affairs for thousands of strangers to read. Thanks, Grandma! A special thanks to Molly O'Neill and Arthur Samuelson, who have helped me in more ways than I can list. Without them this book would not be.

Thanks also to Reed Kelly for his invaluable feedback and friendship through the process, and to Evan McDaniel and Josh Loeb for their help in clarifying certain technical matters. My thanks go as well to my friends at w00w00 and ADM for allowing me to publish something about the inner workings of our cults . . . I mean groups. Henry Korn provided good advice when we needed it bad; thanks, Henry. Olli and Sasha Chanoff read the manuscript at different stages and gave us the benefit of their sub-

Acknowledgments

stantial editorial wisdom. Molly Chanoff helped with the transcription, which made this in part a family affair.

At William Morris we would like to express our gratitude to Owen Laster for his always superb agenting, but even more for his constant encouragement. Thanks also to his assistant Jonathan Pecarsky for seeing us through various trials and tribulations.

Finally, our thanks to Claire Wachtel at William Morrow and to her careful assistant Jennifer Pooley. Claire brings intelligence, vision, and toughness to everything she puts her hand to. She is, very simply, one of the best in the business.

Hacker Cracker

Prologue

Evans was emotionally unstable. He was angry and abusive, and sometimes he raged at people. Everyone worried about it, especially the CEO. Evans was the chief technical officer at Empire Media (not its real name). If he ever decided to wreck them it wouldn't take him more than sixty seconds. One minute at his desktop and he could delete everything on all their systems, with no chance of recovery. They'd have to go to their backups, which were never current, and even then it would take them hours and hours before they'd be up again. They'd lose their weather data, news feeds, and newspaper and magazine updates. For a huge worldwide media company like they were, the losses would be astronomical. Every day he was there the risk got bigger.

But getting rid of him was just as risky; knowing it was coming might drive him over the edge. What they had to do was get him away from his machine long enough to shut him down, which was what my team and I were there for. We needed to change all the root passwords—the ones that gave him the highest level of access to the company's computer systems. We had to find and destroy his back doors—the hidden passages he'd created to get him back in if the front doors were locked. And we had to do it while Evans was in the building, where we could see him. If we tried it while he was away from the office there was always the chance I'd connect up and find him logged on from a personal machine. If that happened, it would be a shoot-out as to who closed down whom.

The plan they came up with went like this: They'd ask Evans to come to the CEO's office, which was in the building across the street from his. At a normal pace, that was a six-minute walk. When he arrived, the CEO would fire him as gently as possible, trying to avoid a blowup. With any luck, that could take fifteen, maybe twenty minutes. After that it would be another six minutes before he could get back to his desk. Meanwhile, as soon as he was safely inside the CEO's office, they were going to flash us the signal to start. From that point we'd have twenty or twenty-five minutes to get into his systems, change the passwords, deny him access, and shut him down permanently. I didn't think it would take more than five.

I felt like the man in black, the unseen avenger brought in to take down the bad guy. I'd been fooling with computers since I was thirteen—seven years back, hacking since I was fourteen, working security since I was fifteen, summers, weekends, and after school. During the day I wore my white hat. At night after work I put my black hat on and went back into the underground.

A dim light filtered into our computer room where I was sitting with George and Das, two UNIX administrators who had been working at Empire as my company's consulting team. George and

Das weren't security people, but they were smart and knowledge-able and we had gone over how we were going to handle this. Everyone had his assignment. Empire Media had multiple connected networks, lots of machines with lots of passwords. But we had divided them up. We were locked and loaded, ready to go the instant we got the word.

In front of our machines a fogged glass wall shielded us from anybody who might wander down the outside corridor. From the ceiling came a steady stream of bad techno music, somebody's weird idea of how to keep us relaxed, as if they thought, *They're technologists, they must like techno.* I tried to put it out of my head. Our machines were already logged on to our targets, windows opened to the respective servers. I was also connected with Evans's personal machine, a powerful UNIX desktop. He was very technical, unlike a lot of chief technical officers—CTOs—who tend to be either theory guys or business types who happen to know something about technology. Even if a CTO is technical at first, an executive career usually takes him away from it. But Evans was good. I knew all about him; I'd been watching him.

Sitting there waiting for the phone call, I tried to put myself inside his head. Angry, hostile, a really bad attitude toward his bosses, at least suspects what's about to hit him. So what's he planning? What would I be planning if I were mad and upset and knew I was about to get it? What devious ways would I use to back door the system so I could let myself in later and wreak havoc?

I ran through my mental checklist of the things we'd done over the last two hours pretending we were just the usual consultants tidying up the systems. The dial-up accounts—did we make sure they belonged to actual people? All authorized, no phonies? The three separate firewalls—anything added secretly to the rules, carving out a secret passage through the gateway? Any other back doors? I knew all of them—the ones available on the internet, the ones not available on the internet, and some on top of that. I knew how you hide them. I knew how you add unauthorized accounts, and how to hide them. All the things a good hacker knows, tricks of the trade.

In the back of my mind the clock is ticking and I'm thinking that the call should come through any minute: "He's in the CEO's office! He's in the CEO's office! GO! GO! GO!" I'm imagining Evans taking his last walk, his rage building, opening the door to the CEO's office, stalking in, wondering if this is what he thinks it is, thinking about the surprises he's got in store for these bastards if they have the balls to try and fire *him*, the booby traps, the bombs he's going to blow them up with. Then, just as I've got this scene going in my mind, suddenly *our* door bursts open and one of the execs sticks his head in, shrieking, "He's on his way out!! He didn't stay there at all! He's on his way out!!!"

It's like a jolt of high voltage. For a second I feel like my mind is short circuited. I can hear Das shouting in his maximum-velocity Indian speed-accent, which I can hardly understand even under normal circumstances, when he's not panicked. He's only been in the country a couple of months and I don't think he's completely tuned in to normal English. It sounds like "WaDIDeesaywa-DIDeesay?" "The guy's left!" I'm yelling. "He's going back to his office. That's what he said. He's left. He's gone. He's out of there. Start changing passwords! George, start changing the passwords." And George is going, "I'm on it, I'm on it already."

By now my fingers are flying over the keyboard. I'm nailing root passwords one after the other. Six minutes at a normal walk, though I know this guy's not walking normally. He's storming. He's so pissed he can't see straight. So how much time does that give us? Three minutes? Four? I can hear Das counting, "One . . . two . . . three . . ." and I know he's ticking off the machines he's supposed to get. George too, "Four . . . five . . . six . . ." I've got mine now—seven, eight, nine. Okay, what else? The backups. We've got to get them too. Okay, you do this, you do this, I'll do that. Checking. Do we have all the roots? All the guest accounts?

Suddenly I'm aware there's activity on Evans's machine. He's back; not just back, he's on. I type in the command that lets me know what he's running: `ps -aux | grep xterm`. I can't actually see what he's doing, but I can see the process identification numbers for everything he's got up. I can see he has five windows open on his terminal, five "xterms," each with its own number. Since I'm logged in as root, which is the highest access, I can kill everything he's got. I can close down his machine remotely too. Shut it off like a light.

I change the root password. On the command line I enter the kill command `kill 1452` for the first xterm. KILL! And I see on my screen `1452 killed`. On his machine the first window closes up and disappears from his monitor. Now he's got four. `kill 1453`. *Shhhp*. Another window closes up. I'm imagining him sitting at his desk, watching his windows close, blinking his eyes. `kill 1454`. *Pchhhtoo*, a third window closes. Staring, understanding now what's happening to him. Knowing there's somebody sitting in a room, probably not too far away, logged on and taking him out. `kill 1455`. The fourth window closes.

Now he's trying to type in a command. We're connected, and he's got root, same as I do, which means he can see me the same way I'm seeing him, he can see my process ID number and do to me exactly what I'm doing to him, terminate my connection just like I'm terminating his. The first thing he's trying to do is change the root password back to another password of his choice. Then he'll run a kill command on the window I'm using. Or maybe he'll just barge ahead and destroy everything he can in the time he's got left. Log in to one of the main servers as root and do an rm-Rf/. "rm" for remove, forward slash for the top-level directory, "Rf" for right now, immediately, without requesting further verification. Forward slash web server, forward slash magazine server, forward slash newspaper server. Deleting everything in the system. No recovery. Serious, massive damage. Wrack and ruin.

That's why it was crucial that we change all the root passwords, and why I can't give him time to change back to a password of his choice. Already I'm typing kill 1456, killing his fifth and last open window. Getting a sudden memory flash from the time a system administrator and I were at war just like this, battling each other for control, except that time I was on the other side.

Now all five of his windows are gone, but he's still logged on and he can just as easily open another window, another xterm. So now it's time for the deathblow, the shot between the eyes. UNIX gives you the lovely ability to do a remote shutdown. I type in init 5. "Five" indicating the level of shutdown, "five" for completely off, total shutdown, as if someone pushed the man's power button. And he's there watching all of his background processes dying, one by one, all the background applications his machine is running. Shutting down this, this, this, this, this, in a mannerly way. He's sitting there watching it all happen, in my imagination he's watching it in horror. And he's seeing messages flash across his screen as each process turns off: SHUTTING DOWN WEB SERVICE, SHUTTING DOWN MAIL SERVICE, SHUTTING DOWN

NEWS SERVICE, SHUTTING DOWN NFS SERVICE. These messages are flying by on his machine, one after another. And he's a technical person. He knows exactly what's happening. And then, when it's finished killing all of the processes safely, the computer itself shuts down. *Shhhhhooom*, his screen goes dead in front of him. And I'm thinking, *Yes! All right! Is this what I love, or what? This is what I live for!*

Bed Stuy, Do or Die

I didn't know my mother was dying, even after I called home and talked to Grandma. For the last two years Mom had been in and out of hospitals, whenever her AIDS got really bad. But she had always recovered. Just a week earlier she had been hospitalized in Virginia, where she was visiting my aunt, which should have meant she'd most likely be okay for a while before the disease came around and hit her again. Maybe I should have been more worried—in Virginia she had almost died. But I talked to her the day she got back and she sounded pretty good. So I wasn't expecting her to have to go back in anytime soon. But when I called home Grandma told me she was at Woodhull Hospital and she was pretty sick. That I should go visit.

On Sunday my uncle Osie picked me up at my place in Harlem in his little red Toyota. As we drove we exchanged a few words, mostly about my new computer security job with one of the city's largest brokerage firms. But it was pretty subdued. We'd both been through a lot of hospital visits and we were psyching ourselves up for this one. I was thinking how much I hate hospitals. I'd rather visit someone in jail ten times than go to a hospital once.

At Woodhull we went in past the security guards and took the elevator up. Walking down the narrow corridor toward the room I saw a guy lounging against the wall with two or three of his friends. Neighborhood thugs, same as the ones I'd been seeing all my life. Woodhull's in the Bedford Stuyvesant section of Brooklyn, next to three different projects, Marcy, Tompkins, and Tilden, where a lot of tough-guy rap artists and big-time drug dealers come from. Home of the famous saying "Bed Stuy, Do or Die." Not far from Grandma's house, where Osie and I both grew up.

Osie and I walked by, almost brushing them—this hallway was a kind of narrow side corridor, not one where you could make a wide detour around someone. Glancing into the open door of the room opposite, I noticed three or four other guys inside, visiting. They all looked like hoods.

On the car ride down from Harlem I had prepared myself to be encouraging. Mom was sick, but I planned to put on a happy face and try to help her make the best of it. The moment we walked into her room I saw I had been fooling myself. She was lying in bed with the sheets tangled around her thin, haggard body. As Osie and I stood there, she kicked feebly at the top sheet, as if she was trying to get it off her. She didn't look quite alive. Her eyes were closed and sunken. She seemed to drift in and out of consciousness. When we came up to the bedside she opened them halfway, but she wasn't looking at us or at anything else in particular either. I was thinking, This isn't really her; she's already gone. Her legs were moving again, trying to kick the sheet off. She was murmuring something in a low, hoarse whisper. When I put my ear down close to her I could hear her saying, "Mommy, Mommy."

I tried to talk to her. "This is Ejovi, Mom. It's Ejovi, your son." She seemed to understand. She asked us to take her to the bathroom. They had her wearing an adult diaper, but she wanted to go to the bathroom. We couldn't, though, she looked on the edge of death. We thought that if we tried to pick her up she would just die in our arms. Seeing her like this was enough to break your heart.

I watched, thinking to myself, *This isn't really my mother I'm looking at. This isn't her.* I was near tears. I always tell myself, *You have to be strong. You can't show weakness. The moment you show weakness is the moment you become vulnerable.* But I could feel the tears coming, so I walked to the entrance of the room, trying to look casual, my back to my mother and my uncle. And I started crying. I held my fisherman's hat to my eyes to wipe the tears, hoping Osie wouldn't notice.

Finally I turned back around and walked toward the bed. I'm sure Osie knew I was crying, though he didn't look at me or say anything. Instead, he had his head down near my mother, talking to her. She seemed to be listening, but she wasn't able to respond. I looked down, then had to turn and walk back to the door. She had been so beautiful, my mother. In high school she was a dancer and an actress. If she hadn't had me when she was sixteen maybe that's what she would have become. When I came back to the bed we talked for a while, Osie and I did. I knew she understood we were there and who we were, but all she could say was that she wanted to go to the bathroom. She was moving her arms and feet, kicking off the sheet we kept putting back. When finally we said we had to leave, she said, "No, no."

Osie is a quiet man, the kind who keeps his emotions to himself. Maybe we both have those genes. Somehow we both got it ingrained in us that there's no point in displaying your feelings, especially low feelings, the kind that will bring other people down. I was thinking, I can't stand seeing my mother like this, in this awful place. I was in a state of shock, numbed. But Osie's grief was coming out in anger.

He was mad and I knew exactly what he was thinking. Osie is thirty-four, fourteen years older than me. Except for when he was away at college, I had lived with him my entire life. He didn't have to tell me how he felt; I knew. He was mad at the people who had given her this disease. Mad at whatever it was that had driven her to drugs. Angry that the sheets were dirty, that the room hadn't been cleaned. That she hadn't been cleaned up.

I knew what was going on with him, but I was lost in my own thoughts. I had dreamed for years about being rich so I could help my family. Buy my grandmother a house. Do everything I could for my mother. But now what could I do? I would have spent every dime I had to get her home. Since she had gotten sick I thought that if I got rich maybe I could even buy a cure for her. Put her on the same treatment Magic Johnson is on. She'd be okay. She'd have leather coats, minks. She'd have a private nurse and her own doctor. My fantasies. If only she could wait long enough, until I could get the money together. I had a couple thousand in the bank already, and I had an excellent job now, one I'd just started. With my salary, combined with Osie's, the two of us might be able to do it, if only she would wait. But it didn't take a genius to see that she wasn't waiting for anyone. I said good-bye to her. I leaned down and kissed her. I didn't think whether I would ever see her again. I refused to think about it. So I just kissed her, and said good-bye.

Osie and I left, each of us wrapped in our own misery, walking down the narrow corridor. He was a step in front of me, talking softly but intensely, to himself mainly, about complaining to the nurses, except it was Sunday and we had only seen a couple of nurses on our way in, and no doctors. Sunday, and the place was dead. He was talking about how he was going to come back tomorrow and make sure they started taking care of her right.

Up ahead of us I saw the same hoods lounging against the wall. The one I had first noticed earlier was half sitting, half leaning on a covered ventilator with his legs sprawled out. Some of the others who were inside the room before were lounging out there with him.

When we got up to them Osie stepped over the guy's outstretched leg without seeming to notice him. Then I stepped over, mumbling, "Excuse me." Next thing, one of the thugs lets out with, "Hey, Howdy Doody," meaning Osie, I thought, or it could have been me. Probably Osie, though, who was never the most fashionable dresser you'd meet. Everyone else in my neighborhood might have been thinking about what kind of statement they were making with their sneakers and how their pants hung. But things like that never made an impression on Osie. He had always been immersed in his books, studying first for his bachelor's, then for his master's in mathematics, then for his CFA. Osie was born and raised in Brooklyn, but you couldn't tell it from his clothes any more than from his attitude, which was always shy and reserved.

The "Howdy Doody" didn't bother me. My mind wasn't on it. So I just kept walking, even when the hoods started hooting with laughter, probably over how clever the remark was, and maybe also because Osie and I obviously weren't going to stand up to them. Osie was so consumed by his anger he didn't seem to have noticed either the remark or the laughter. But suddenly, a hundred feet down the hall, he spun around and let out this loud braying laugh, *Haah haah aaah aaah aaaah,* like he was mocking their laughter, except it just didn't come out quite right.

Haaah haaah aaah aaah aaaah! It had a kind of strangled sound, like someone choking. I thought, *What the hell are you doing?* There were six or seven thugs back there, and Osie isn't an aggressive guy. I seriously doubted he'd had a fight since he was a teenager. *Okay,* I'm thinking, *you had your laugh, now let's just keep moving.* The elevators weren't that far in front of us. Maybe we could get there before these guys decided they had to chase us down.

We started walking again. But now the thug down the hall was yelling, "What you sayin', huh? You wanna do something?" And Osie turned back, saying, "Yeah, you want some? C'mon, let's go," and started walking back toward them.

I grabbed him. "Osie, come on, man. Let's just go. We don't have time for this. Come on." Down the hall the thugs were getting up, more of them coming out of the room to see what was up. The short guy, the one who had started with the Howdy Doody remark, was already walking down the hallway toward us. And as he came he was cracking at Osie, "You want some? You wanna do somethin'? Huh?" And they were coming faster now, seven of them, including one being pushed in a wheelchair.

There was a mop leaning against the windowsill right where we were. I grabbed it, thinking, Okay, let's see how much of this kung fu staff fighting really works. In the years I spent practicing kung fu I hadn't done much with the staff, but I knew a few moves and thought I could improvise some others. At least enough to keep them away from us.

The corridor wasn't more than six or seven feet wide. Maybe I could hold them back with the mop handle while someone was getting security. But they were coming and when they saw me grab the mop they started with, "The guy's grabbing a mop? He's gonna use a mop? C'mon, motherfucker, you gonna use a mop on us? Let's see what you're going to do with a mop!" I started twisting the mop head out for dear life, I needed that mop handle for my kung fu staff. I was twisting and twisting but it wasn't coming out. By now Osie and I were backing up side by side. I was desperately trying to unscrew the head until it finally dawned on me that this was one of those safety mops, where the head doesn't come off, exactly so people like me won't be able to use it as a weapon.

By now Osie had his fists up, in some kind of boxing stance that looked like it was from the 1890s. He was yelling, "You want some of this? Huh? You want some?" And now the short guy was right up in his face, snarling, "You four-eyed motherfucker," while I was trying to smooth it over, calm it down. "C'mon now, we don't want any problems. We just want to get on the elevator and go our separate ways. You do your thing, we'll do ours, that's it. See what I'm sayin'?" I muscled Osie back a couple of feet to where he was out of range, but he was shaking with anger.

My uncle Osie is brave. I know that. And strong. Once when I was twelve he took me away from my stepfather, who is a big, heavy powerhouse. But Osie's only slightly taller than me, 5'8" or so, and thin, no match at all for these guys. I wanted to put a stop to this, just get us to the elevator and out of there. But by now the insults this guy was hurling at Osie were getting me hot too. Not many things push me to the point of getting physical, but disrespect is one of them. That's something ingrained from childhood. And the amount of disrespect this guy was piling on was making me really want to pound him. Worse, I knew he was only talking this kind of trash because he had six guys behind him, which made me even madder. Inside, I was steaming. Every instinct was screaming for me to unload on this guy, just take him out. *Man, I was thinking, if you were alone, or if there were just two of you, I would drop you so fast you wouldn't have time to blink.*

But I didn't. In studying martial arts over the years, I learned that being a fighter isn't just about throwing punches, it's about thinking, and dealing with situations correctly. When I first started learning, one of the fundamental things we were taught was that if the guy's got a knife or some other weapon, or if there's more than one guy, you use a special weapon of your own against that. It's called *running*. And I knew this was exactly one of those situations where if you can get out, you get out. Fast as you can. But Osie was maddened about my mother, and now it was all spilling over onto these punks.

And right in the middle of this, just as I was thinking how to wind it down, Osie hauled off and punched the short guy right in the face. *Boom.* In that split second everything slowed down. I thought to myself, *My God, what did he just do?* and I knew there was no turning back. Once the first punch is thrown, there is no more talking. That's it. All I could do now was start swinging. What was done was done. Now it was time to rumble.

I was standing a little to the side, separating Osie and the short guy, when Osie threw the punch. When that happened I turned to the guy and banged him with a one-two combination, right, left. I knew we were about to get the crap beat out of us and I just wanted to get in as many hits as I could. So *boom*, right, *boom*, left. The short guy didn't go down, but he was in a state of shock. He didn't think Osie was going to hit him, and he certainly wasn't expecting it from me. He reeled back, stumbled, then caught himself. He was holding his mouth, saying, "What the fuck?" And then they were all rushing us.

Instead of staying with me, Osie split into the side corridor with his back up against the wall, two or three guys coming at him. I could hear him fighting as I tried to keep the rest of them off me, thinking, *Don't get grabbed, don't get grabbed. Just don't let them get you down where they can start stomping you.*

Except for the short guy they were big, the littlest probably had forty or fifty pounds on both of us. And the first guy coming at me was the biggest of them all, wide shoulders and thick body, Spanish-looking. I was pedaling backward with my hands up. I jabbed him hard and felt my back up against the wall. The first thing they teach you in martial arts is to get your back against the wall. A light-skinned guy was coming from my left, trying to decide how to get at me. The Spanish guy was right in front of me now with his hands up. My hands were up too, getting ready for whoever was going to grab for me. But they were hesitating, so I took a step in and *whoom*, I kicked the Spanish guy in the thigh with all my might, a roundhouse kick, the hardest I can throw. I know from experience what that feels like. It feels like somebody slammed you with a baseball bat. His leg crumpled and he froze. I saw the fear and bewilderment in his eyes. I heard myself screaming, "C'mon, c'mon now, c'mon!" I was ready to go in and pound him until he cried for mercy. He probably came after me because I was trying to be the peacemaker, so he thought I must be afraid, that I'd be weak, an easy target. I could feel my anger overflowing. I was in a rage, ready to unload on him with everything I had. Like being in the ring, when you've got your opponent dazed. You can see that he's weak, you've hit him hard, you've hurt him. That's when you go in for the kill.

But just before I went after him my head suddenly cleared, and I came to my senses. Two others were coming at me from the left. The wheelchair guy was ahead in the corridor. If they cornered me there was nowhere to run. So I kept my hands up and I figured, as much as I wanted to hurt the big guy, I had to get out of this situation or they were going to get me, wrestle me down, and stomp me to death. And there was no way out except to run through them. Because no matter how hard I hit these guys, even if I knocked one or two of them down, it only takes one to grab you. And once you're grabbed, that's it. If there's more of them than there are of you, you're gone.

I threw my hands up and the light-skinned guy came toward me. I faked and ran past him, taking a swing—not really hoping to hit him, just wanting to keep him back so I could get free. My burst carried me through and I was about to go toward Osie, who was fighting three guys now. But the guys I had just gotten past were coming at me and Osie looked all right, for now, anyway. He was on his feet throwing windmills like he was possessed. And I was thinking, *I can run and get help myself or stay and fight side by side with Osie, those are my choices.* Staying was going to be a losing battle, with all of them on us in close quarters. But if I ran to the nurses' station to get help, then sprinted back, at least we'd know that someone was coming, we'd know we wouldn't be in the situation where these guys could beat on us until they got tired. So I decided to run the fifty yards down the corridor to the station. I ran yelling, "Get security! Get security, already!" Two nurses were there staring at me with these dumbfounded looks on their faces, and my brain was screaming, *What is going through your minds? Call security!* And I was shouting at the top of my lungs, "Call the damned security!" But they just stared. They seemed paralyzed. Okay, I thought, I told them, that's the best I can do, and I spun around and raced down the hall to get back to Osie.

As I was sprinting back I saw the thugs leaving, running in the other direction, pushing the guy in the wheelchair. When I got there Osie was on one knee, getting up from the floor. He had a lump on the side of his eye, but other than that he seemed all right. Part of me was thinking, *Why did I run to get help?* What if one of them had pulled a knife and stabbed Osie? How would I ever have lived with myself? Osie was alone, and you don't leave your partner alone even for a moment. The right thing would have been to stay and go down with him, fighting back-to-back. That's the story I would like to tell in the neighborhood, or tell my children twenty years from now when they ask me about my young manhood. But I hadn't done that. This is what I'd been training for all these years, a situation where it was them versus me, or them versus us. And I failed the test. I should have stayed and gone down in a hail of punches and kicks. That's what I should have done.

Osie picked his glasses up off the floor, and his hat. "How do I look?" he asked. "Not too bad," I said. "Just a shiner." We went to the window and saw the thugs already running down the street below, pushing the guy in the wheelchair. "How come you didn't stay with me?" said Osie. I shrugged it off. "You look fine," I said. "Just fine."

But I didn't feel fine. I felt terrible. I felt like I'd been sucked right back in, no matter how far I thought I might have come from those streets I grew up on. My mother was dying in that room back down the hallway and I was in such a state of pain and confusion I could hardly think. But that made no difference. You can be in any state you want and the streets will just suck you back regardless. You can't get away. Worse, I felt ashamed and angry. I knew I didn't do the right thing. After all the training, all the fighting, here was that feeling of helplessness again, that feeling of having no control. That same awful feeling I'd been trying to get away from my whole life.

2

Be Prepared

Every weekend when I was little the man I thought of as my father picked me up from Grandma's house and drove me to his place in another part of Brooklyn. He was African and spoke with a strange accent. He told me he was from the Kokori tribe in Nigeria and that his father was a chief. My name, Ejovi, also came from Nigeria. It was an African prince's name. At the end of the weekend when he brought me back he'd always give me some money and I would buy Icees and candy for all the kids on the block.

I thought of him as my father, but my mother told me he wasn't my real father. My real father, my biological father, was a man I met briefly when I was five or six but have little memory of, a man who died of kidney failure when I was thirteen. I think my Nigerian father also knew he wasn't my natural father, at least in the back of his mind, though how far he believed it I wasn't sure.

After all, he had given me my first name, Ejovi, and my last name too, Nuwere. That was confusing. If Francis Nuwere from Nigeria was not my real father, then why did I carry his name, and why did he call me his son?

But though it might have been a little confusing, at age six or seven, it didn't make much difference to me. Grandma's house, where I grew up, was full of people who seemed like family though some weren't really related to us. The house, actually the second-floor apartment of a three-story building on Pulaski Street in Brooklyn, was a happy, noisy place so crowded that you could hardly keep up with all the comings and goings. In addition to my grandmother there was at one time or another my mother; my aunt Donna; my half aunt Sadel; my cousins Brian, Geneva, and Stacy; my "grandfather"—that is to say, Grandma's friend Doc; my uncle Osie; my uncle E-Bone; my kind-of uncle Big Will; my little brother Jevon; and me. That was one lineup, but the inhabitants were always in a state of flux. All of us living in a three-bedroom apartment, which seemed a lot smaller than it probably was.

With so many people in the place other relatives and friends more or less came and went at will. Very rarely did anyone bother to knock. We knew everyone on the block and they knew us. The building, especially, was like one big family; I probably ate almost as often at my friend Manuelito's upstairs as I did in my own.

Everyone around called my grandmother "Grandma," and there was never any doubt about who was in charge—she was. My grandmother was beautiful, five feet six inches or so, thin and vivacious, with a lovely smile. With her there was no fear, even when she was disciplining us kids, which she did by whacking us with a plastic baseball bat (the bat's still there, right next to her place on the sofa). When we needed correction she'd chase us down and bop us, which didn't actually hurt but did make a loud whacking slap sound. When we got a little older it became a joke. We'd make believe we were crying, then go into the corner and start laughing. But we had to watch out. If we had done something serious, Grandma would turn the disciplining over to Big Will, which was a different story.

Grandma slept in the living room with Doc on the foldout sofa while the rest of us shared the bedrooms. During the day, when she was home cooking and cleaning, the sofa would be up and she'd have her music on—Al Green, Teddy Pendergrass, especially Marvin Gaye. She'd play the old songs on her record player and dance with me, swaying and waving her arms to the rhythm. I loved dancing with Grandma, even if the music was a little weird. Those singers of hers were old. Marvin Gaye didn't sound anything like Run-D.M.C. or LL Cool J. Or Michael Jackson, whom I idolized and was planning to be exactly like when I grew up. To prepare for that I practiced his dance moves until I had them down to what I considered perfection. Then I'd perform for Grandma, Doc, and the others, moonwalking across the room like I already was Michael.

Grandma's house was always loud and busy, which was just the way I liked it. You could never be bored. There was always someone to play with, someone to talk to, always someone around if I was unhappy or afraid. And everyone loved one another. The family was tight, and the apartment always seemed full of laughter. Every day there were people stopping by, partly for the free meals. Grandma would never refuse to "feed a mouth," as she put it. Or they'd come over for cards. When that happened everyone would sit around the table while I drifted from one player to another peeking at their hands. There was always alcohol on the table and thick smoke hanging in the air from so many cigarettes. But the gambling was for pennies, just an excuse for friends to get together and hang out.

One of the regulars was Sug, who always wore plaid shirts with polyester pants and a Kangol hat. "It's Sug," he'd rasp, his voice deep and bluesy. "I just came down to see what's going on." This tended to be right around lunch- or dinnertime. I'd run and jump on his lap like he was another uncle. He'd do a little hat trick for me. He'd raise his eyebrows and his hat would go up and down on his head. Then he'd do the finger trick, pulling his finger off and putting it back. He was an alcoholic—even at that age I noticed that he drank a lot. He seemed to have a pint of Bacardi next to him at all times.

Sug and Doc were friends and Doc drank a lot too. Tall, six three or four, with a shaved-bald head, Doc looked like a basketball player. Doc taught me about doing business and making money. He'd ask if I wanted to help him earn some cash, then we'd go to a liquor store where they were selling discounted half bottles of whiskey by the carton. I'd wait outside and he'd come out with one, sometimes two boxes. We'd push them home in a shopping cart, carry the cartons upstairs, then wrap each bottle separately in newspaper. When we got that done we'd put them back in the boxes and push the cart from Brooklyn all the way to Manhattan, across the Brooklyn Bridge at breakneck speed, as fast as my legs would go. In downtown Manhattan we'd go into a park and sell the bottles to the alcoholics there for a dollar fifty or a dollar seventy-five, maybe twice what Doc had paid.

Eventually Grandma kicked him out of the house, because Doc not only drank, but was a drug addict. It was obvious to me. I was an early expert at spotting addicts. I would go to the bathroom after him and see blood drops in the sink, or bloody tissue. I never actually caught him shooting up, but Grandma did. He still had the rubber band around his arm. He always had rubber bands on his wrists, the telltale sign of a drug addict. Grandma caught him and was screaming at him to get out of the house. He was yelling, "I didn't do anything, I didn't do anything!" But after that he was gone and that was the last I saw of him.

When I was eight or nine I became aware that my mother was having problems. For a glee club performance I had once, maybe in first or second grade, I needed shoes. Grandma, who was on welfare, gave her money for them and we went to the store. We were looking and she said, "Do you like these shoes?" "Yes," I said, "those are nice." For some reason we didn't buy them, but when we got outside the mall she gave them to me. After that she would take me with her to malls where there were big department stores. I'd keep lookout while she shoplifted shoes, clothes, or whatever else she could. Even now, whenever I go into a department store I have the feeling that somebody's watching me. Actually, they usually *are* watching me. I fit the description. But I look even more suspicious because I have this lookout mentality. I didn't see my mother's stealing as that horrible a thing. She was doing it to get us the necessities, clothing and things, so we'd have what we needed when the new school year came around. Even if it wasn't exactly right I knew she was doing the best she could to take care of us and be a mother to us.

But other things were more disturbing. We'd go out to neighborhoods that were really bad; dangerous places where we'd meet shady people. I never said anything, but I knew these were drug addicts, and the guys she talked to were dealers. When you see things often enough you start putting them together and figuring out what's going on, no matter how young you are. I saw it but I never talked about it, not with my friends, not with Jevon, not with anybody. Those were the secrets I kept inside.

Grandma must have been worried sick over what was happening with her daughter, though I didn't know about that. What I did know was that there'd be occasional explosions of anger. One time Grandma was holding money for my mother and my mother decided she should get it back. Grandma had told me where she'd hidden it, and told me not to tell. But when my mother asked, I couldn't help myself. When she went to get it a big argument erupted. Grandma told her that the money was for when the kids needed something. My mother didn't agree. Grandma was screaming, "You can't just take this money! You asked me to hold it for you for the kids. You can't do this!" "No!" my mother was yelling. "It's my money. I can decide what I want to do with my money and my children. They're *my* children!" She grabbed me by the hand, hoisted Jevon up, and ran out of the house with us. That night we went to a shelter. Jevon and I waited outside for a long time, playing in the cold and shivering because we hadn't taken jackets along, until finally they let us go in. Inside there were other kids with their parents. Some had been burned out of their houses, some were just plain homeless. All of them looked as lost and confused as we were.

I have a vivid memory of playing in that cold street, but what happened after that is jumbled. We may have slept in the shelter that night, maybe for a couple of nights. Eventually we ended up at a welfare hotel in Sheepshead Bay. For some reason, after we moved into the hotel my mother wasn't around during the day, though she came back at night. A friend of hers who was also staying in the hotel would look after us and help us cook. We'd take our food to the hotel kitchen. If there was a chicken I'd take the hair off of it and do the gravy. I'd mix it up like I thought you were supposed to but it would come out whitish and gloppy, and not very good-tasting. But we ate it anyway.

Before this I had rarely seen white people, except for teachers and cops and once in a while the Mormon missionaries who would come through the block smiling at everybody like they were taking a stroll through the park. The whole neighborhood would stop and stare at them in wonderment. But in Sheepshead Bay, everyone was white. Other than in the welfare hotel, you couldn't find a black person. Once my mother and I were walking past a big white-colored building with a lot of white people going in and out. I asked her what it was, and she said, "Oh, that's the Ku Klux Klan building." I didn't know if that was a joke or what; maybe it *was* the Ku Klux Klan building.

Like everything else in Sheepshead Bay, the school I started going to was all white. I had never met a white kid my age, and the kids out there had never known any black children. I was the only one in the entire school. I didn't know what to expect, but it worked out really well. Right away I realized I had suddenly become popular. I was the cool black kid and the other kids wanted to hang out with me.

The whole scene out there in Sheepshead Bay was cool. The hotel was nice. I had friends in the school who liked me. And after school I went to play baseball with kids in the neighborhood. I had a beach, or at least rocks, where I could go and hang out. The whole ocean was my backyard. I was as happy as I'd ever been. But before I could get really settled in we were gone again. A month or two and we were out of there.

Not that I minded going back to Grandma's; I had plenty of friends at home, too. There was my next-door neighbor John and his brothers and sisters; Carlos and Donald, who lived across the street; and Ito from upstairs. Plus B'lal and his sister, Shakina, whose names were African, like mine was.

On weekends some of us would meet up and go to the free lunch at the elementary school, and in the summer we'd go during the week too. Sometimes Ito's mother would walk us over, or John's mother would. Sometimes John's sister would come with us and sometimes John's whole family would come. Weekends and summers that's where we'd mainly hang out, playing ball in the schoolyard. Except on Sundays when the Yogi Bear Bus would come to get us.

Yogi Bear was a church bus that went through Brooklyn picking up children for a kind of free religious day camp. We called it the Yogi Bear Bus because Yogi was their mascot and there were pictures of him all over the bus. We'd pile on and all the kids would be singing, "He's got the whole world, in his hands / He's got the whole wide wor-ruld, in his hands / He's got the whole world, in his hands / He's got the whole world in his hands." A big bus full of kids careening through Brooklyn, everyone singing at the top of their lungs: "If you're happy and you know it, clap your hands [*clap, clap*]. If you're happy and you know it, clap your hands [*clap, clap*]." The bus was jumping from all the kids singing and bouncing around like Superballs.

Finally we'd pull up at this big building like a warehouse, where they had a makeshift gym and a stage. We'd sit in front of the stage and the counselors would ask questions about religion. "Who was the son of God?" Everyone would raise his or her hand, and if you got the right answer you'd go up onstage to get a prize, or even better, they'd let you "jump for money" and actually give you some depending on how far you could jump.

"Who was the mother of Jesus?"

All the hands are up and everyone's yelling, "Me! Me!"

"Tanisha?"

"It's Mary, Mary."

"Yeeeees! Come up onto the stage and get your prize."

Then they had plays: The Devil versus Jesus, or Good versus Evil. The counselors would take the parts while the kids were the audience. And all the plays would present some moral question. The counselor onstage would say, "Oh, she just found a wallet. What should she do? Should she give it back? Or keep it?" And half of us would be screaming, "Keep it! Keep it!" while the other half were like, "Give it back! Give it back!"

It was great; I loved it. We all did. I looked forward to going every Sunday. We'd stand in front of our building waiting for the bus, Jevon and I. All year round. My grandmother would wave at us from the window, saying good-bye. The bus driver knew my grandmother by then, like he knew almost all the parents. Those Yogi Bear counselors were some of the few white people who penetrated the neighborhood. They might have just thought they were giving us a fun way to learn about God, and they did do that.

Eventually, when I got to be seven or eight, I grew out of Yogi Bear. The guy would drive up and say, "Ejovi, you're not coming on the bus today?" I'd say, "No, I'm just going to chill over here."

"Okay. But when are you going to come back?"

"Next time. Next time I'll be there."

But I had grown out of it. It wasn't cool anymore. After a certain age the kids who went on the bus got made fun of. Just like the kids who went to the free lunch program got made fun of. So I had to stop that too. As much as I enjoyed them, I had to quit those things. If I didn't, I knew I'd just get myself teased and beat up.

Instead we walked around the neighborhood looking for things to get into, which weren't hard to find. Our block was littered with abandoned buildings and vacant lots with lots of interesting things in them. If we found a mattress in a lot that was big fun. We'd jump on the mattress, wrestle on it, do flips on it. God knows where that mattress might have come from or what it had seen, but who cared? We'd just pull it into the middle of the sidewalk and have fun for as many days as we could until the garbage people came and took it away.

We'd go through abandoned buildings to see what was in there and to find cool places to hang. If a building had been burned out you might really find unusual things—toys, kitchen utensils, clothes. We'd look for anything intact. Once we found a piano. We'd play tag and hide-and-go-seek, racing wildly through the lots and the buildings. In one of the lots there was an apple tree and in the summer we'd stage wars, running behind the houses, jumping over fences, throwing the crab apples at each other.

When there were no apples available, or even when there were, we'd make bean shooters: take the plastic bottle from a twenty-five-cent juice drink, cut the bottle at the neck, then fix a balloon on one end with a rubber band. Put a stone or a dry bean in the opening, pull the balloon back and you had a pretty good home-made slingshot. From my stoop I could hit a window across the street at full force and even break it, which I did a couple of times. We'd buy a ninety-nine-cent bag of beans and split them, then we'd have wars, running around and shooting each other. Hours and hours of fun for a dollar. We'd run through the abandoned buildings shooting each other, like cops and robbers, except we didn't call it that. We just ran and shot. Played see if you could survive with the beans coming at you at a hundred miles an hour. If one of them hit you, you would have a good little welt for days.

Somewhere else maybe there were kids whose parents took them to the park or to ball games. But the playgrounds were too far from where we lived and our parents limited us to just our block, so we had to make do with what we had. An outsider might have wrung his hands and said, "Oh, those poor kids. It's so dangerous." But for us it was our normal life, it was what we did. Lots and vacant buildings were our playground. We played tackle football in those lots, our side of the block against the other side. We played baseball with a bottle for one base, a brick for another, a piece of wood for a third. We'd hit the ball, run the bases, sometimes slide in. It's a wonder half of us didn't die of tetanus, given the scrapes and cuts. But it didn't seem to do us much harm.

There was real danger in the neighborhood too, violent danger, though when we were young none of it seemed to be directed at us, so we only became aware of it by degrees. We often heard shooting, but it wasn't coming in our direction, and as long as you weren't in the crossfire there didn't seem to be much to worry about. Instead, we made it into a game. My brother, my friends, and I would try to guess the kind of gun we heard. "Yo, that was a thirty-eight." "Yeah, and that was a nine." "Tec Nine, Tec Nine." When there was some kind of holiday, people would be shooting all the time and we'd have a bonanza. The next day we'd go through the neighborhood picking up shell casings. See who could get the biggest, a .45 or a .357, a nine millimeter. We'd walk down the street scanning the ground and somebody would say, "Oh, look, I got a nine, I got a nine." At the end of the day we'd count them up and see who had the most.

Drugs were there too, drugs were everywhere. You knew who the drug dealers were. They were your friends, they were the cool guys who hung in front of the grocery store. They were the people you looked up to, the teenagers we wanted to be like. Until it dawned on me one day that these were the same guys who were selling drugs to my mother. And when I understood that, my admiration for them dissolved and was replaced in my heart by hatred.

Over time I developed a revulsion toward the crackheads and heroin junkies I'd see on the street. One of the skills I picked up from living in my neighborhood was the ability to spot an addict a mile away. It didn't matter if he was wearing a sharp suit or a dirty jogging outfit; you could tell from the way they walked, the limp, the eyes, the skin tone. You see it so much you can spot them anywhere. They were not in control of their lives, any more than the alcoholics looking to score a drink. That was the opposite of the way I wanted to be. I thought I'd rather be dead than live like that.

And sometimes we did see them dead. Like the time I was walking down DeKalb with my mother, coming back from the Chinese restaurant. Across the street I saw a crowd gathering around an ambulance parked in front of a vacant lot. In the lot was an abandoned white truck that had been there so long it was like a part of the landscape. The truck's back was open and facing the street, so I could see directly inside, where two EMTs were standing over a body. I heard someone in the crowd say something about how somebody had OD'd. There was a sheet over the body, but the dead guy's arm was sticking up, bent at the elbow. It was still up when they put him on a stretcher and took him into the ambulance.

I wasn't shocked in the least, even though you didn't get to see a dead body every day. I just thought, *Oh, another dead drug addict.* I was repulsed, but I was scared too. Kind of the same way I was scared every time I went into my friend John's house, which was right next door to ours. John's parents did drugs, and they didn't hide it. Often they'd send John out to make a buy for them, and one time when I was there they sent me with him. "Go to Angel around the corner and get some." John gave Angel the money and Angel gave him a foil packet. But even Angel wasn't that comfortable. He said, "You gotta tell your parents to stop sending you." "All right, all right," John said, "I'll tell them." John was his parents' mailman.

John and I grew up together. We were tight, but his house was a house of horrors. For starts, it didn't have a toilet. I don't know if their building was condemned or if the toilet was broken and they just didn't fix it or what. But they would come out with a bucket and pour the excrement into this shallow hole in the street in front of their house, where it would build up. My friends and I would horse around and try to push each other into it, that pit of crap. Inside their house it was filthy and dark; it smelled. It was a dangerous, scary place. My grandmother's house was so crowded sometimes you had to wait for the bathroom, but compared to people throwing their crap on the sidewalk, I was living the high life.

That was a dysfunctional family. John had a little brother

named Willie who was the terror from hell. He struck fear in every person on the block, child and adult, and he was only five or six years old. His parents just didn't care about the things he did, they thought it was amusing. He would grab sticks with rusty nails in them. He'd chase us around the block—not playing, really trying to rake us or stab us. He'd pick up a brick when you weren't looking and heave it at you.

Everyone was terrified of that kid, but the whole family was erratic. Once my uncle Eric had an argument with John's older brother, who was about Eric's age. To resolve it John's mother, Cheryl, pulled out a shotgun and came to our stoop. She pointed the shotgun at my grandmother and the others on the stoop, saying, "Y'all want to bring trouble, huh? You wanna bring trouble?" Waving the shotgun.

John's family didn't just do drugs, they often sold them too, mostly out of the basement of our building. The addicts would be lined up through the hallway, down the front stairs and outside. You had to get around them to go upstairs. Sometimes the cops would raid and the flashing lights and bullhorns would set the block on edge. The family was also famous for having an arsenal in their house. Whenever we wanted to see a gun, we went to his basement, where we knew they were hidden. We would go down there and look at them, examine them. There were so many I didn't know if they weren't also dealing guns. John would take us into the dark basement—actually, they kept the whole house dark, as if they had something against light. He would hold the guns for us and let us look at them, including the black sawed-off I'd seen his mother hefting out on the street.

John was cool. He had Atari, and he had those guns, so it was hard to stay away. But going over there gave me a thrill of horror.

It frightened the shit out of me, kind of like the way I would be frightened by a horror movie. But it didn't make me think seriously about dying. That didn't happen until I was ten. And it didn't happen at John's, but on the stoop of the abandoned building on the other side of his house.

That stoop was a favorite place for me and my brother Jevon to sit and pass the time. One sunny summer day we were hanging there as usual, talking about nothing, watching the cars go by, when I noticed three guys strolling down the block, older teenagers. One of them, a guy I'd seen around the neighborhood a lot, wore a blue bandanna. For some reason I got a sudden chill, a kind of premonition of trouble. I watched them walking, then noticed a little crowd was gathering down the block. I said to Jevon, "Wonder what's going on there?" A minute later we heard *pow*, and the same three guys came running past us, laughing and joking while they're running, like something funny just happened.

I stood up and saw the crowd was getting bigger. And there was my grandmother walking down the street to see what happened. I went to follow her but she said, "No, stay here. I'll be right back."

A couple of minutes later she came back with a look of disbelief on her face. She walked right past Jevon and me to our building, where people were gathering on the steps. "They shot Blue in the head," she said. "His brains were on the sidewalk. He was moving, trying to breathe, but he's dying for sure."

I didn't go down to look. My grandmother's words stuck in my head. "His brains were on the sidewalk." "They shot Blue in the head." Blue was older than me, but I knew him well. Everyone called him Blue because of his blue eyes. He was a good friend of Uncle Will and was up in our house all the time. I imagined him lying on the sidewalk. I could see it. And those guys running down the street laughing. And at that moment I knew that your life could go just like that. In an instant. One moment you could be playing in the street and the next you could be dead, not movies dead, really dead, with your brains splattered on the sidewalk. I was ten years old. The runners were teenagers. I saw their faces. I remembered them. They were all familiar, but one I had seen a lot. And a couple of months later he was back in the neighborhood, walking down the same street. I thought, *Your life can go at any time. You have to be prepared.*

3

Chk-Chk-Boom

B ut death was only one of my fears. Like all my friends, I had other things to be afraid of. Like the Decepticons.

The Decepticons—the Decepts—were a gang of older teenagers, real tough guys who stuck people up and robbed them. A bunch of them lived on Throop Avenue, a block and a half away. They were all over the Tompkins and Sumner projects up in that direction. We knew never to go there. If you go that far into some unknown place, you're bound to get beat up and robbed, so we never ventured to that point. But at the same time, the Decepts didn't come down to Pulaski, where they would have run into our local dealers and thugs. So we didn't actually see them around that much. We just knew to stay out of their way.

There were also the younger ones, though, the baby Decepticons, kids about my age. And I did see them a lot. The game arcade was in a grocery store a couple of blocks from where I lived, in a kind of neutral area, and they came down there to play the games and rob people of quarters. They always traveled in packs and carried weapons, razors and such. The baby Decepts would come into the arcade, maybe five or six of them. I'd recognize them from having seen them walking around the neighborhood. Friends of mine had gotten punched and robbed, and if they came into the arcade and I was by myself I knew I was in trouble. I was away from my block, which meant I was in a risk area. So if I saw them coming I'd take evasive action and do what I could to avoid them. Finally my friends and I from around Pulaski and Tompkins decided to start our own gang so we could protect ourselves. The Decepticons had gotten their name from a cartoon show, *Transformers,* where the Decepticons were the evil villains. The good guys on the show were the Autobots, the Decepticons's enemies. So that's what we decided to call our gang, the Autobots.

It didn't take long before we started up a full-fledged war against the baby Decepticons. Our idea wasn't to wait around until they came onto our block. We started traveling in packs and made it known that we were out to get them. So now they were the ones who were afraid. No one had ever challenged them before.

There was always a lot of fighting in the neighborhood, that was just a fact of life. But now we started having mass rumbles. Pulaski between Throop and Tompkins was the middle ground between us and them, and that's where we'd meet. Our gang would go up there and find some of the baby Decepts and the two groups would start waling away at each other, always turning it into a running battle. People would be charging around, ducking behind cars, punching somebody, then taking off either to get away or to chase somebody else down and punch him. One time we chased them all the way down Throop toward the projects. I'd run after them, catch one of them. We'd throw our hands up, circle around, throw a couple of punches, then start moving away if a friend came to help the other guy.

Even though these were wild brawls, for some reason they never got totally out of hand. People might get beat up and hurt—once I was clotheslined while I was chasing one of them and barely managed to get back to Pulaski—but no weapons ever came out. So it actually got to be more fun than anything else. And after a while the two groups of kids became closer, even though we were still separated because we were from different parts of the neighborhood. But after those fights we began to respect each other, so we got to be more like friendly rivals than worst enemies.

Most of the baby Decepts were the older Decepts' little brothers, which could spell trouble. If some baby Decept got hurt during one of our fights, there was always a chance the big Decepts might come down the street looking. That happened to my friend Carlos, who had beaten up a baby Decept and later that day found himself up against the kid's older brother, who must have been about sixteen. Carlos was big for his age, but he was only eleven and he wasn't about to face up to a sixteen-year-old. But Carlos's older brother and sister wouldn't let him back down. "If you don't fight him," they said, "we're going to fuck you up ourselves. We'll beat the shit out of you."

So Carlos had no choice. He was crying, wiping his tears away, but he had to do it. We were in a circle, me and our friends, the other guy's friends, all of us waiting to see if someone was going to jump in, which would have started a massive free-for-all. But nobody did. Carlos and the big kid threw up their hands and to everyone's surprise, Carlos started waling on the big kid. That was pretty good. That was Bed Stuy all the way. Bed Stuy, Do or Die.

Bed Stuy, Do or Die—that meant we were the toughest, tough as they came. We were always ready to fight. Always ready to rumble. It gave me a sense of pride, a sense of belonging to something. No matter where I went in the five boroughs everyone knew Bed Stuy, Do or Die.

Not that I ever actually went to the other boroughs. What I knew about them was mainly from rap songs. "Bronx is the craziest / Bed Stuy's the dangerousest." But I had no idea what the other boroughs might be like, except for Manhattan, which I had seen from having gone there with Doc to sell whiskey.

But when we got a little older, my friends and I would venture into Manhattan once in a while. That wasn't something you did by yourself. The train station was right outside of the Marcy projects, which was definitely not a place where you wanted to get caught alone. Or even with too few partners, which happened to me once when Carlos, Fila, and I decided to chance a day trip.

We covered the four blocks to the Myrtle-Willoughby station carefully, eyes out for trouble. But none showed until we got down into the station itself. Down there on the platform we were looking at a subway map on the wall trying to decide where we'd get off when suddenly we realized we were being surrounded. When we turned around there were eight or ten young thugs looking at us like we were some kind of prey animals and they were the predators. I was wearing a knit ski cap with earflaps, the kind that had just become popular after some rappers on MTV started wearing them. Everyone wanted one and I had managed to get a bright orange one. As they backed us up against the wall one of them said, "How much did you pay for that hat?"

"I don't know. Ten dollars."

"Can I see it? Can I see it?"

"Okay," I say. I'm thinking, What am I going to do, say no? So I give the guy my hat.

"Oh," he says, looking it over, "this is nice."

Another one takes it from him and says, "This is mine now."

I'm saying, "Nah, nah. Gimmee back my hat, man." But I'm scared. There were way too many of these guys.

"Nah, man. You're dead. You're dead on this." Meaning it wasn't mine anymore. It was his.

So that was it. I lost my hat. We were lucky that was all that happened. That was why we had to walk together in groups; the more in the group, the better.

It was a strange life. Schizophrenic. On one hand I was the happiest kid. I had everything I wanted—friends in the neighborhood, always something to do, a loving family. Whenever I came home there was hot food or something in the fridge to eat. My grandma and my aunts would cook and there was always more than enough for everyone. Most of the adults in the family were working—maybe not in good career jobs, but they were bringing home paychecks. We weren't rich, but we weren't the poorest either. I had no complaints, especially when I looked at my friends.

We weren't living in the projects like some of them, where once you're there it's hard to see anything else. We weren't living in the dangerous crap house next door. I knew who my mother was, even if she wasn't around nearly as much as I wanted her to be. I might not have had a father, but I had father figures. I had my Nigerian stepfather. I had my uncles: Osie, who was in college; Eric—E-Bone—who had a big rep on the streets; Will, who held a good, steady job. I had my aunts, my cousins, my little brother.

I'd look at my friends, and so often the only ones they had in their lives were their mothers. Or they were being abused by their fathers. Or their living conditions were scary. Some of them were deprived, no question about it. But it never once occurred to me that I might be.

I even found a way to make good money. When a new super-market opened down the block from our house I went to check it out and noticed a kid packing bags next to a checkout counter. As I watched I saw that after he'd bag someone's groceries they'd give him a tip. Next to another checkout line there was the same kind of table, but no one was bagging there, so I asked the cashier if I could help. She said I could. The baggers weren't employees, they just worked for tips. As long as I did a good job it was fine with her. So I started packing, and people started tipping me. They were giving me money for doing what I had been doing for my grandmother and my mother for years. It was amazing. That day I must have made ten dollars. The next afternoon I made twenty-five, and I figured that if I came really early, like when the store opened, I could probably make a fortune. Since it was summer-time and school was out, there was no reason I couldn't be there right at nine o'clock.

I worked the rest of the summer bagging groceries, and when school started up I worked afternoons and evenings. I got to be very fast and efficient. I learned how to take the tips out of my bucket every five minutes or so, leaving only a few quarters in there so that customers would know to leave a tip, but they wouldn't see how much I had already made. The money built up like crazy. People might only leave fifty cents, but the store was always busy and a couple of hundred shoppers might go through my line a day. I was making fifty dollars, a hundred on big days like weekends or the week before Thanksgiving. I was hooked up with capitalism. At the end of the day my pockets were so loaded with change I could barely walk.

So on one hand my life was normal: school, friends, family, a job. No complaints. But on the other, I had the feeling that things were closing in. Danger and trouble were everywhere, and as I got older they started to weigh down on me more and more. There were burdens and circumstances I had no idea how to escape from. I began to feel they were pressing in on me from all sides.

■ ■ ■

On the corner of Tompkins and Hart there were some broken-down chairs and old crates that made it one of our favorite spots to hang, despite the dealers who operated just down the block. We're there one day, talking and horsing around as usual, when this old guy in a dirty trench coat comes around the corner and stops right in front of us. We know who he is—a creaky old neighborhood wino whose son is one of the big local drug guys.

"You seen my son?" he says. His voice is old and raspy. He's slurring his words. "No," we say. "We ain't seen your son, man." We're not paying him much attention. He's kind of disturbing us, standing there and staring at us like that. But he's just an old derelict wino.

"Naw," he says. "I think you seen my son."

"Whatever, man."

We're ignoring him, hoping he'll go away and stop staring at us with his old alkie eyes.

"I think you seen my son," he says again, while he's opening his trench coat. "Where the hell's my son?" And as he says it he pulls out from under his coat a big black tommy gun, like Eliot Ness's, with a big round magazine, and the black hole of the barrel is pointed right at us. We scattered so fast he didn't have time to pull the trigger, though I don't know if he was planning to or not. I was gone like Road Runner, my heart pumping so hard I thought it might explode. I couldn't believe it. Pointing a gun at us like that. I'm flying, listening for the burst of bullets, thinking, Are we all going to be on the ten o'clock news tonight?

When I was younger, real danger never seemed to come close. It was more like a game. But it wasn't a game anymore. You could be killed dead. Some were, and you could be too. You might not see the other side of eleven, easy.

Ito's uncle upstairs wasn't playing games either. We didn't know any details, just that he was selling drugs and he had gotten into some beef with dealers from the Marcy projects. Those were guys you didn't want to have problems with, especially if they knew where you lived.

Usually if there was gunfire at night from somewhere in the distance it didn't wake you up, it was just part of the background noise. But one night I woke up to a shot that was real close. *Boom!* As I came awake I heard more. *Boom, boom, boom, boom, boom,* then BOOOOOOHM—a shotgun. Coming from directly above us. *Boom, boom, boom, boom,* then BOOOOOOHM, the shotgun again. Like we were in the middle of a war, except it was all happening right in front of our building. Down in the street people were shooting up at the apartment over us, directly above our window. And upstairs, someone—Ito's uncle, we figured—was blasting back at them with the shotgun, *chk-chk* BOOOOOOHM. *Chk-chk* BOOOOOOOOOOHM.

Jevon and I hit the ground, which is what you did automatically when you heard shots nearby. We lay on the floor listening to the gunfight as it flared up, then wound down to some random shots, then stopped. When it was over we got up and spent an hour or so talking about what had just happened, until we calmed down enough to go back to sleep.

Ito's uncle led a dangerous life. After the gun battle we knew the project gangsters were looking for him for real. Troubles that usually went on in the street were coming up to where we lived, and it happened again not long afterward. This time we were in the living room, and heard a commotion on the stairs. People running and shouting, then a hammering on our door and Ito's uncle shouting, "Let me in! Let me in!" My aunt Sadel went to open the door, but before she could the shooting started—four shots, one right after the other, *boom, boom, boom, boom*. Then more commotion, getting farther off, though huddled down in the living room we couldn't tell exactly in which direction it was going. When it finally seemed safe to look we opened the door, not knowing if we were going to find a body or what. No one was there, the hallway and stairwell were empty. But four big bullet holes were clustered in a tight pattern in the middle of our door, at just about eye level. We speculated about why they didn't get him. He was at the door pounding to get in when we heard the shots, so from the placement of holes it looked like a miracle they didn't hit him. People in the building are still guessing about where he might have been and, judging from the holes (which are still there in Grandma's door), where the shooters were.

That wasn't the end. Sometime later somebody blew a shotgun blast through his door, just missing his baby, who was asleep in the next room. And eventually the project boys did catch up with him. They shot him twice—two different times, though neither time in the building. Amazingly, he recovered from his wounds both times, though for a while I'd see him walking around gingerly with help from a cane.

Ito's uncle wasn't the only one in the building with trouble. My own uncle Eric, Osie's younger brother, had more than his share. Eric was really well known in the neighborhood, mainly by E-Bone, his street name. Everyone was like, Oh yeah, E-Bone's little nephew. So I already got a certain amount of respect just by being related to him.

Eric and Osie were only a couple of years apart and they were close growing up, which was strange in a way. They both had the family quiet gene, but other than that you could hardly find two more different people. Osie was always studying, always into his books. Practically the only time he would come out of his room was to get something to eat and maybe talk for a minute before going back to work. It seemed like every time I'd go in there he'd be reading and listening to his jazz music. I couldn't understand why someone would listen to jazz when there was such good hip-hop to listen to. I would go into his room and he'd have ten books spread out around him, working with some high-tech calculator. Listening to that jazz.

Eric was just the opposite. He was a throw-down type of guy. Always ready to fight, always up to something, one of the guys everyone in the neighborhood knew, and knew not to mess with. Eric treated me like a younger brother, same as Osie did. We played around, but he never took me anywhere, I think because most of the places he went were troublesome. He'd come home, eat, hang out. Maybe he'd do some break dancing for us. He always had girlfriends, was just the cool uncle. I never saw his bad side; all I knew was his reputation.

When I was growing up I didn't understand to what extent Eric was in trouble. I had the idea that he did bad things once in a while, though exactly what they were I wasn't sure. But as I got older I began to figure out that he probably didn't get the gold chain or the new jacket he came home with on his own. He probably took them from someone.

Whatever he was doing, they weren't things he wanted the family to know about. Maybe Osie and Will might have known more. Maybe he didn't say anything to me because he thought I might go tell Grandma. But I think he did try hard to keep his street life separate. There were times when it came crashing down, though. Lots of times.

Eric left for jail when I was thirteen, but long before that I was aware the cops were after him. They would come knocking on our door. Cops, or sheriff's deputies with warrants, saying, "Have you seen Eric?" Or we'd look out the window and there would be a cop car parked across the street, and we knew they were waiting for him.

Once they almost caught him. I was sitting on the sofa with Jevon and a bunch of cousins watching TV when Eric came running in with his best friend, June Bug, on his heels. A minute later the police were pounding on the door. "Open the door! Open it!" Pounding. *Boom. Boom.* We opened the door.

"Where's Eric at? Where's he at?"

"We don't know. We don't know."

"How about the other guy, the guy who was with him?"

By now Eric's hiding. June Bug's on the sofa with me, my brother, and my cousins. June Bug's like, "No, no. I'm just sitting here with my nephews and nieces, Officers. We're not doing anything." And Grandma's saying, "What's going on, Officers? What's this about?"

They started searching through the apartment. There was one door that was locked and they didn't have a search warrant, so they couldn't open that up. But they searched through all the other rooms. The cops were screaming at my grandmother, "We can lock you up for this! We can take your grandchildren away. How about you shut up and tell us what we want to know?" Very disrespectful, I thought. Very rude. And to my grandmother, who I thought of as this old woman—though she probably wasn't older than her late forties.

I had no idea what was going on. I just knew one thing—to be quiet if I didn't have to say anything and to say the right thing if I did. And June Bug is going, "These are my nephews, right?" And we're like, "Yeah, Officer, he's our uncle. He's our uncle." Eventually they left. Later Eric came out of the locked room and went out the back fire escape. So he got away that time. Another time, though, he wasn't so lucky.

He was in the house one afternoon hanging out with a bunch of us when suddenly someone was trying to kick down the door. *Wham! Wham! Wham!* We all went into a panic. God only knew who might be trying to break in. Maybe it was the guys who were after Ito's uncle and they had mistaken our apartment for his. I grabbed the phone and called 911. I'm yelling, "Help us, help us! Someone's trying to break into our house!" Everyone's screaming. Outside big voices are shouting, "Open the door! Open the fuckin' door!" Kicking it, kicking it. I'm on the phone crying, "Please, please, send the police. Send the police. Send someone. Please."

And while I'm screaming into the phone the door smashes open and it *is* the police. Looking for Eric. They start tearing the place apart and they find Eric hiding in the living room closet. They're dragging him out, yelling, "C'mon! You're coming with us." And Eric's yelling, "No, no, no! Ma. Ma. Tell them to let me go. Ma. Tell them." I'm crying. My brother's crying. They're taking Eric away. We're sobbing, "Leave him alone. Leave him alone."

They're dragging him, handcuffed now, and he's still screaming, "Ma, tell them I didn't do anything! Tell them to let me go. Ma!" And one of the police goes, "Shut up!" and *bam*, he hits him across the face with his walkie-talkie. Eric's eye is bleeding. They're pushing him down the steps. Out the window I see them shoving him into the car, hitting him.

Grandma would try to protect Eric from the police if she could, but she had her own confrontations with him. She wasn't happy with his ways. When she was really angry, like when he would come home drunk, or try to sneak some woman in, she'd smack his face and call him a bum, shout at him in the hallway just outside my room, where I'd pretend to be asleep, lying there agitated but thinking that the best thing was just to shut it out and mind my own business.

■　■　■

Most of my life, certainly through my childhood, I was a really quiet kid, reserved, like Osie and Eric. I was quiet because I was always thinking about my life, about what was happening, about my situation and how I could solve whatever problems were bothering me. I thought too much. I still do; I obsess. When I was younger every small thing troubled me, *really* troubled me. Even now I'm an overthinker. Give me a situation and I'll hypothesize a hundred scenarios. I'll worry it to death.

When I was in sixth grade, I started becoming more aware of the bad things going on around me. All the violence. All the drugs.

So much went on in the neighborhood that I didn't understand. Why were there so many drugs? Why did the dealers keep selling to mothers and fathers walking with their children? I couldn't stand seeing mothers holding their kids' hands and buying drugs. Seeing families of friends of mine buying drugs and looking so awful. Seeing my friends' mothers high on heroin when I went to visit. Knowing when they'd just finished smoking crack. I'd look at the crackheads walking by my stoop, feeling disgust, thinking, *How can they be so weak? How can they do this to themselves?* There was so much I couldn't comprehend, so many things that were wrong.

I felt helpless. There wasn't anything I could do about any of this; I just had no control over the situation. There were so many troubling questions I needed answers to, but who could I talk to? Grandma was there, but she had a million things to do and so many people to take care of. My mother was the one I was closest to, the one I loved with all my heart. She was the one I needed to talk to. But now, when I really needed her, she wasn't around. Instead, she was in jail, on Rikers Island, though I don't remember exactly why. The Christmas when I was twelve we went to visit her. By then I was getting used to visiting people in jail. I knew that before you went in you had to put your things in a locker, which cost fifty cents. Then you had to walk through a metal detector. Then you had to be searched by one of the corrections officers. Then you went and sat down at a table to wait for the person you were visiting. I would sit at the table, look across the hall, and see my mother in the holding area, where the prisoners were brought before they were allowed into the visiting facility. We would wave, waiting for her to come and see us.

My mother looked so beautiful that Christmas. When she came

in she gave us presents—presents the prison system made available to all the kids who were visiting. But presents or not, she was still wearing an orange jumpsuit. It was still prison.

While she was in jail we'd get letters from her; they were apologies, explaining how she felt, and how things would be when she got out. Then she'd be released. She'd come back out and say, "I'm never going back to jail. I'm going to stay here with you kids. We're going to get a house. We'll get our own place. We'll be independent. We'll do so many things. We'll be one perfect, happy family together." But then a couple of months later she'd be back in jail again, my mother, who I loved like my life. I believed every word she said, every time. So when she didn't keep her promises, I was just ruined. I was ruined every single time.

The fact that my mother was where she was, that I couldn't see her whenever I wanted, affected me probably more than all the rest of it combined. In my usual way, I obsessed about it. She was the one I needed, the only one I wanted. But she wasn't there. She was in jail, where she told me she wouldn't be anymore.

I must have had a hundred things on my mind, problems I had in school, in the neighborhood, with my friends. But without my mother, I had no one to talk to about them. So they kept building up inside me. It wasn't that the problems were important, most of them were just little things. But those little things built up and built up. Just the normal problems of growing up—at least, the normal problems of growing up in a neighborhood like mine. And the hugest problem of them all, the one I really couldn't deal with, that my mother was not there.

The result was that all I could do was think. My grandmother was great, a mother figure. But I didn't need a mother figure; I already had a mother, but an absent one. I was also without a father. A kid like that is going to either become a bad kid, or a troubled kid, either turn it outward or inward. I kept it in. I went in the troubled direction.

As these small things began to accumulate inside me my outlook on life started changing. I started seeing things in a different way. I looked at the people in my neighborhood, the addicts and alcoholics, the gangsters and thugs. I saw all these violent incidents happening. The value of life seemed so much less to me than it had before. Obviously none of these people put any value on their lives. And slowly the feeling crept up on me—no one here values his life, why should I value mine? What did I have to live for if I didn't have the things I truly needed, like my mother, whom I loved, even though she put me through pain sometimes. Who had gone through hard times but had always done her best to be there with me. But now she wasn't, so what else was there? It began to seem like the answer to that question was *nothing*. Maybe there was nothing else.

I had been a happy kid before, but now I felt sad all the time. I began falling out with my friends in the neighborhood. Nothing big, really: I had a fight with one of my old friends. I had always played with him, and now we weren't friends anymore. That was a small thing; two weeks from then we would have been friends again. I went to a school where kids teased each other all the time, a normal thing. "Hey, that Ejovi, he can't get any girls." Taking turns about me not getting some girl to like me. But in my state of mind that small thing, that teasing, was more than enough to set me off. These were the guys I went to school with every day, my partners, the guys I hung with, the ones I faced the world with.

I began having a feeling I couldn't shake—that my glass was full, and I mean this was a *huge* glass. One guy makes a joke, a drop in the glass. Someone else makes a joke, another drop in the glass. Another joke, another drop. Until one of those drops made the glass overflow. It overwhelmed me. I didn't have a mother. I didn't have a father. And now I didn't have any friends anymore either. And finally I said to myself, What's the sense in continuing? Why not just get it over with?

■　■　■

It was in school that I decided to kill myself. At lunchtime. In the middle of the period. The idea, which had been building inside me, suddenly was too strong to fight anymore. A wave of sadness overwhelmed me, leaving me without a single hope. I felt it close in on me and there just didn't seem to be anything else I could do. I knew the time had come to put an end to it.

I left the lunchroom and walked through the empty halls. I passed by the office, where the principal and the other office people were. I walked up to the balcony that hung over the stairwell and I looked down. It was a long, long way, a couple of flights before you'd hit the ground. I stood there for a few minutes and thought about what it would be like to do it. How easy it would be, just to finish it here. No one would care if I jumped. If I died, no one would care. So what was the sense of sticking around?

I climbed over the wooden balcony railing, which had support bars painted gray. I grabbed onto the gray bars and looked down at where I was about to go. For a minute I stood there in a kind of trance, just thinking about letting go. Then I heard someone come out of the office. Looking up, I saw it was one of the secretaries.

Her eyes widened and then she was running toward me. At that moment I let go. And just as I did she reached down and grabbed me by one hand. She was holding on and crying, "No, no! Don't do it, what are you doing? Help, help me! Someone help me!"

I looked down, and I realized in that moment how real death is. The fact that I was looking at death, and that this person was so afraid for me, jolted me back. And I decided, dangling there, that I didn't want to die. I didn't want to fall. I started crying, and she was crying even more. Then more people came running out of the office. She pulled me up and I grabbed the rail from the bottom. And I started pulling myself up, still crying. She was screaming, "Oh my God! Oh my God!" as a couple of the other office people helped her pull me over the railing.

She was saying, "Why did you want to do that? Why?" She was hysterical. There was nothing I could say. They took me to the office and called the police and the ambulance.

When the police came they told me they were going to have to take me to the hospital. But I didn't want to go to the hospital, I wanted to go home. Why should I go to the hospital? "No, no," I'm saying, shaking my head. "No!" So they grabbed me, a huge cop grabbed me. The ambulance people were standing there waiting. "Come on," said the cop. "Let's go."

"No," I said. "I don't want to go. I don't want to go. I want to go home." I jerked myself free.

"No, you have to go to the hospital." The big cop tried to grab me again, but I pulled away. Then another cop grabbed me and I was fighting hard against both of them. The secretary who had saved me was crying again, saying, "Just go with them, go with them. It's going to be okay. It'll be okay." And I was screaming,

"No! I want to go home!" pulling away from the cops, trying to kick them, hit them. Then two more cops came in and I was in a fury. It took all of them to get me out of that room in handcuffs.

They took me out. They put me into the ambulance and cuffed me to the stretcher. The doors slammed closed and the ambulance took off. I screamed and fought the whole way, trying to get loose.

4

The Knife

ven while I was tearing at the cuffs I was aware that my grandmother was in the ambulance with me, telling me over and over
to calm down, to relax, saying, "It'll be okay, Jovi. It'll be okay."
Then they were wheeling me into emergency and the rage just
went out of me. I could see the people looking at me, people in the
waiting room, wondering what had happened, what was wrong
with me. The cops were hovering over me, doctors, the ambulance
people. I wasn't fighting anymore; in fact, I was starting to enjoy
it. I felt like a movie star, everyone looking to see if I was okay,
feeling concerned. I felt special. Because I *was* special. I wasn't
Ejovi the sixth grader. I wasn't Ejovi the kid. I was a special person who needed special attention. I liked that.

They were talking to my grandmother, telling her they wanted
to put me in the psychiatric unit for a while, just until I could calm
down some. Then they were wheeling me through the hospital or

maybe taking me to another building, I couldn't tell. Things were beginning to cloud up.

Next thing I remember clearly I was in the psych ward. There wasn't much to it, just two corridors of bedrooms, a nurses' station that looked into both corridors, and a kind of recreation/meeting area with a TV set, where families could come and visit. Overall it didn't seem too bad. Maybe the psych ward wasn't where you would choose to be, but I could easily imagine worse places.

There were about twenty of us in the unit, all kids, me being the youngest. Some of them acted funny, but I don't think any of them were psychotic. Maybe emotionally unstable, like they were saying I was, or maybe they had behavior problems that landed them there. I had the idea that many of them had done bad things and were given the choice of going either to Juvenile Hall or the psych ward. But nobody seemed too violent. Mostly they weren't much different from anybody else.

There was one guy in there who was a Decepticon, a really cool guy about sixteen who took me under his wing. He taught me a couple of Decepticon handshakes and showed me the secret Decepticon transformation they do when they're fighting—"transforming," they called it. Basically this was a set of robotic-type movements where you dip and turn, moving your hands and body to distract the person you're fighting. Then *boom*, you let him have it.

Another guy there was just slow. He talked slow, his gestures were slow, he walked in a slow-motion shuffle. He'd sit there and hang out with us sometimes, but he was too drugged up to really get involved. Then there was this huge, overweight, mentally retarded guy. Childlike. He was older, a man really, but they couldn't put him with the adult psych cases; they would probably have hurt him. So there he was, with us in the children's ward.

Once when I went to the bathroom the Decepticon and the slow guy followed me in. Before I knew what was happening they pushed me into the shower stall where the retarded man was standing buck naked taking a shower. I glanced down and it seemed like his penis was partially cut off, as if it was decapitated. I looked up at him and he was staring at me with this strange expression on his face. I started screaming in terror, "*AAAAAHHHHH*, get me out of here!" but the guys were holding the door closed, trapping me in. Finally they let go and I burst out, running down the hallway while they laughed their heads off. Then the mentally retarded guy comes out of the bathroom smiling and laughing himself, as if he thought it was a great joke, too.

In the ward we slept two to a room in bedrooms that had no doors, except that I wasn't sleeping much. Everybody else in the place was out like a light but I couldn't seem to fall asleep. After lying in bed for hours I'd go to the nurses' station and ask for a sleeping pill, but I'm sure they were giving me sugar pills because they never seemed to help. Our room was all the way at the end of one of the corridors, right next to where the night security guard sat, so I'd often spend the hours talking to him. He was a pretty cool guy. Among other things he studied karate, and in the course of talking to each other every night he began showing me some basic karate moves, my first introduction to real martial arts. Finally, around three or four in the morning, I'd feel myself getting drowsy and go off to bed.

Aside from the fact that I couldn't sleep, being on the ward wasn't a bad experience. One of my aunts worked in the adult psych unit and she stopped by regularly to see how I was doing, so I had daily contact with the family. Then there were the different kinds of therapy. Each day I'd spend time talking to a counselor, and there was an art therapy class where we drew and painted. I was given some kind of pills, too; Prozac, they told me. I also had treatments where they would attach sensors to my head, then bring in a machine that flashed random patterns of light in front of my eyes. I never quite knew what this was about. I'd ask and they'd say, "Well, we want to see how you react to the blinking lights." But why or exactly what they were looking for I never knew. When I wasn't getting light treatment or art therapy or counseling I'd go back to hanging out with the other guys. It was cool, like a gang of friends. More like camp than anything else.

I was enjoying it. I could actually feel myself calming down. When they told me after a month or so that I was ready to go home I was happy, but I was unhappy too in a way. The psych ward was such an easy place to be.

When I left the hospital, my state of depression seemed to have gone away. If I had been able to articulate it at the time I would have said it was over. But what I didn't know then is that depression is a disease you never really get rid of. It's recurrent. I was to have bouts of depression all through my teenage years. It's always there, so in that way I never recovered. I just dealt with it, then and later. I still do. At the time, though, I thought it was over; whatever had happened to me seemed to be gone, probably with the help of the Prozac. But it wasn't, really. Underneath I was still in turmoil.

■   ■   ■

While I was in the hospital the family was talking about what might be best for me after I got out. My Nigerian father, or step-father, Francis Nuwere, thought it would be good to give me a change of scenery. The school year was about over, so he suggested that I stay with him and his family for a month or two. Not a permanent thing, just for a period of time so we could see how it might work out. Grandma agreed. The move would separate me from the things I had been part of, and possibly from some of the problems that may have contributed to what I had tried to do.

So I went to stay with him; his wife; my little stepbrother, Kona; and their new daughter. Francis's wife was in the service and the family now lived in housing at Fort Hamilton, an army base at the far end of Brooklyn. Fort Hamilton was only a train ride from Grandma's house, but in some ways it was like another world.

Francis's house was like a condo, small but comfortable, and the family was comfortable too. They treated me like one of their children and even though I hadn't spent as much time with them in recent years as I had when I was little, they made me feel at home. And right from the start I found friends to hang out with. First was a Mexican kid who lived above me, whose father coached the base soccer team. Then I met Billy, a kid from Texas who said he was heir to a huge fortune his grandfather had left him, that he was going to inherit when he was eighteen. He spoke with a major Texas accent, but I hardly noticed. It was just, "I'm from Texas"; "Great, I'm from Brooklyn." Billy was pretty much a loner and I was the new kid, so it was easy for us to hook up. We were both looking for someone to hang with.

So now there were three of us: the Mexican kid, Billy, and me; four when Kona tagged along. At Fort Hamilton there was lots to do. There was a recreational center with a bowling alley, basketball courts, and a ramp for skateboarding, which I took up immediately. There was a day camp we went to, and a video arcade at the PX, where we could also buy the newest toys with our allowances. Mine was ten dollars a week cash money, which made me feel like a rich man.

I played on the baseball team, hung with my friends, and watched the military stuff that was going on all the time. Twenty-one-gun salutes, helicopters coming down and taking off, military police with their white helmets and armbands. I saw guns everywhere I looked, which wasn't exactly new, except these were in holsters. I wasn't feeling bad about what had put me in the hospital; in fact, I didn't dwell on it at all. There was so much to do that I didn't even think about it.

Before long our group had expanded a little, and now included a pretty girl who started hanging out with me. I would go to my friends' houses and their parents would yell, "Ejovi's here!" And they'd come running to the door.

I'd never had anything like this in my life. At home my friend John's little brother used to throw bricks at me and chase me with a lead pipe. If I went somewhere off my block I might be robbed. Here I could go anyplace I wanted, as long as I was in by dark. I even had a girl who liked me, and I liked her. Jasmine. We kissed a couple of times, but mainly we talked a lot. She had dark blond hair and a beautiful smile. I thought I was in paradise. I had left Bed Stuy for greener pastures.

Not that Bed Stuy had left me. One day my little brother Kona came home crying. When I asked what happened he told me some kid had hit him and said that next time he was going to beat him up. "C'mon," I said. "Show me the kid." When Kona pointed him out, I said, "Hey, were you messin' with my little brother?" and *bam*, I punched him in the eye. He fell down on the ground crying. Later his parents came knocking on our door with their kid, who now had a black eye, saying I had beaten up their son. But I didn't really get into trouble. My brother explained that the kid had been picking on him and I was just sticking up for him.

The next day, a lot of kids were hanging around; it was the Fourth of July celebration. My girlfriend and I were talking and I said, "You want to see something?" I called the kid, "Hey, c'mere, c'mere, c'mere." The kid looked up as if he hoped I was talking to someone else, but he came. I said, "Hey, show her your black eye." The kid showed her, turning his face so she could get a good look. "Look, here," he said, almost proud of having such a great shiner. She turned to me. "Why'd you do that?" "Oh, he was messin' with my little brother," I said, being a big man, "so I had to teach him a lesson."

One day me and Billy, the Texan, were rummaging through the basement of my building and Billy had a great idea. At home his father had a crowbar he could get hold of. We could pop open the locks on the basement storage rooms and see what was in them. That sounded like fun, so he got the crowbar, popped the lock on one of the bins, and opened the door. Inside were bicycles, dartboards, and a bunch of other interesting stuff that people keep in storage rooms.

We didn't take anything out, but it was fun to see what people had. So for a week we popped open storage rooms and looked inside. We even popped his own family's storage room, where we found his father's .38 revolver. "Hey, look," he said, "my father's gun." "Yeah," I said, "that's cool."

But later Billy started telling other kids, including the Mexican kid and Jasmine, what we were doing. Now we had a whole little group to play *Mission Impossible* with. We were on an army base, so we pretended we were army guys. We'd stand with our backs pressed up against a wall waiting for the right moment, and when no one was looking we'd run down the ramp leading to a basement. The basements were damp and narrow, with different corridors, like an enemy base might have. And somebody would say, "Okay, there's a storeroom here," and *pop*, we'd snap the lock.

We only took the whole group once, but that was a big day. That day we went to three buildings and popped lots of storage rooms. Everyone was like, "Oh, wow. Look what I got here. I got a water gun here. Oh, look at this bike, I got a cool bike here."

Then someone came up with the smart idea of taking the bike from this storage room. We spirited it to another building, moving from wall to wall and bush to bush—*Mission Impossible* style—and put it in an empty locker. I didn't actually have any interest in the bike, but it was fun that we got something we could use, a cool bike we could all ride anytime we wanted.

As I later found out, one of the lockers we broke into that day belonged to Jasmine's parents, though she didn't know that. And that evening they mentioned to her that their locker had been broken into and she told them everything. She was being honest, I guess. Where I came from, you didn't do things like that. We had been popping locks for a while now and neither Billy nor I would ever think of telling an adult. Billy understood these things. But Jasmine didn't.

So instead of clamming up she told her parents who was involved. She named names. The Mexican kid's parents found out, and suddenly I was no longer the kid to hang out with anymore. Not only that, the Mexican kid's parents went to the military police. And the MPs said, "You know what? We've been getting a lot of reports of lockers being broken into."

When the Mexican kid's parents came to my house and told my

stepfather I was involved, he asked me, "Did you do this?" And I said, "No, I didn't." I knew I would get in trouble if I admitted it, right? So of course I denied it. Whether Francis believed me or not I didn't know, but he was angry. I had tarnished his good name, and we had to go down to the military police for an interview.

The military police were not courteous to him. It was almost as if they were treating him as a suspect.

"What country are you from?"

"Nigeria."

"What part?"

"The northern part."

"What village?"

"It's only a small village."

"Just tell me the village. What's the village?"

They talked to me too, accusing me of robbing a storage room. I told them I had gone down to the locker rooms with everyone else, but that I hadn't stolen anything. Which was true. All I had done was look through people's stuff. The whole time they were questioning me I was thinking that if the kids had just said, "Let's not tell anyone about this," nothing would have happened. But these kids weren't raised in the right neighborhood. They didn't have the correct ethics to deal with this kind of situation. If this had involved my friends at home, everyone would have denied it and they would have had nothing to hold us on. That would have been it.

As it was, this whole thing was an extreme humiliation for my stepfather. He was a very strong man who needed to keep face. In Nigerian culture, that's a priority, and because of what I'd done, he'd been disgraced. He had been personally disrespected by the police, and his son was now known on the base as a bad child.

This was a major situation. The atmosphere in the apartment turned tense; I could feel his anger ready to burst out. "You're not to do anything," he said. "No phone calls, no going out. You are just going to stay here in this room." He was smoldering, which made me extremely uncomfortable. I was beginning to have a really bad feeling about it.

I didn't like the way this was going at all. The police were accusing me of robbery. I wasn't getting any support from my friends, because I had no more friends. I wasn't allowed to see them and they weren't allowed to see me. And the only family I'd known for the past month was turning on me. I had no one, and my stepfather seemed like he was a minute from waling away on me. Who knew what was going to happen to me now? This was bad. Really bad. This wasn't a place I wanted to be in anymore.

So I told my father that I wanted to go back to my grand-mother's house. And he said, "No. You're not leaving. You're not going anywhere until this whole situation is dealt with."

I waited for a time when I was alone and I called Grandma. I told her what was going on, that I thought my stepfather was going to do something to me. I told her they were accusing me of robbing these places, which I hadn't. My voice began to shake and suddenly I was crying, pleading on the phone, "Could you please come and get me, Grandma? Please come and get me!"

Later that same day my uncle Osie knocked at the door. "I'm here to take Ejovi home," he said to my stepfather. And my step-father said, "No, you're not. Ejovi's not leaving. He's staying here with me."

"No, he's coming with me," Osie said. "I came here to get him and I'm taking him. Ejovi, get your things and let's go."

"No," said my stepfather again. "He's not leaving."

"Look," said Osie, "you can't hold him here. He's not your real son anyway. He's Barry's son"—Barry was my real father, the man I had met so briefly once whose name I knew but whom I barely remembered. "He's not yours!"

"I don't care." My stepfather was getting angry now. "He's not leaving here!"

I already had my things packed, so I grabbed my bags and headed toward the door. But my stepfather saw me coming and started closing it. "No, no," he said. "You're not leaving!" And Osie slammed the door open and pinned my stepfather, who's a big guy, behind it. And suddenly they were pushing and shoving each other. The door swung one way, then the other. They were shoving back and forth, pushing and heaving in a wild commotion.

By now I was completely panicked, sobbing and shouting, "Stop! Stop! Stop fighting!" I didn't know what to do, how to make them stop or how I could get out. Instead of thinking, my instincts took over, and my first instinct was to grab a weapon, thinking, This is a big fight; a weapon's going to help. Where I was raised, if you can't beat someone, you get a weapon. You pick up a brick and hit the guy over the head. You pick up a knife and you cut him. My mind was yelling at me, Get a weapon, get a weapon! And in a split second I was in the kitchen and I had a big knife in my hand. I ran toward Osie shouting, "Here! Here!" And I handed the knife to him.

The moment he saw this my stepfather stopped struggling and a look of despair and fatigue came into his eyes. He just stood still and looked at me. Later, much later, he told me that at that moment, seeing his son pick up a weapon on him, he felt that I had killed him already. He told me that in Nigeria the father is the most important person in the family. In Nigeria a son will die before letting anything happen to his father. He'll lay down his life to protect his father. And I had picked up a knife on my father. I had committed the ultimate sin.

To me it wasn't that way. It wasn't clear cut at all. I was a kid in a situation he didn't understand. Who was afraid for himself and afraid for his uncle. Who was simply afraid, and had no idea what he should do. And out of panic, not a rational decision to get a knife and harm someone, I reflexively grabbed a knife. I wasn't thinking, *Hurt someone;* I was thinking, *A knife will get us out of here. A knife will keep him back.* And I just did it. I gave the knife to my uncle.

Osie was holding the door open, now with the knife in his hand. I ran out. Then Osie ran out, and threw the knife down in the hallway.

On the train home, Osie said, "Why did you give me that knife? That was stupid, we didn't need a knife."

"I didn't know what to do," I said. "I was afraid."

I didn't speak with my stepfather again until my mother passed away eight years later. After that afternoon at his house on the base I never heard from him. But if I had looked around me more carefully when I was outside playing on weekends, I might have seen his car sitting on the corner of Tompkins and Pulaski, the driver staring out the window trying to catch a glimpse of me as I ran around the street with my friends. And if I had been able to see through space later, when I was rarely home, I would have seen the same car with the same driver straining to pick me out among the teenagers hanging on our block.

5

I Have a Dream

When school started that September, Grandma still thought it would be a good idea to keep me as far away from neighborhood problems as she could. So instead of going to Junior High School 117 in Bed Stuy, which had a bad reputation, I found myself living with my aunt Donna in Flatbush and going to school a couple of blocks from her house.

Flatbush was new to me. It was a part of Brooklyn I had never spent much time in, so I didn't know the neighborhood and I didn't have any friends there, though after the first couple of days I picked up a couple of new buddies who lived in my aunt's apartment house and went to the same school. But being a new kid in a new place, I was watching out for what kind of trouble might come my way. Which, of course, it didn't take long to do.

One day when I was at lunch I realized that this kid was staring at me, giving me the look. Grilling me. He's looking at me and I'm looking back at him, and he's like, "What do you want to do?" "Whatever," I tell him. "What do you want to do?" Then one of his friends pipes up, "You want to fight him?" "Yeah," I say, "okay," and I stand up, getting myself ready to go. "All right," the kid says, "I'll see you after school."

The rest of the school day I'm hearing bits and pieces of talk that this isn't going to be a fair fight, that when I leave school the kid and his friends are going to jump me. So when the last bell rang, instead of leaving I went to the school office and told the principal what had happened. "These guys are going to jump me," I said. "They're going to beat the crap out of me as soon as I leave. You need to do something." "Wait a couple of minutes," he said. "I'll walk you out."

When the principal finished up what he was doing he walked me to the front door. But as we were going down the steps outside he saw an old student of his and said, "Okay, you can go home now. I'm going to talk to this fellow here." So instead of walking me to the corner or the two blocks to my aunt's house, he stopped at the front door. Which would have left me alone except that my two new friends were waiting for me. We're walking and I'm saying, "I can't believe this happened, leaving me like that. How hard would it have been for him to walk me a block or two?"

By now we were at the corner and there was a knot of kids across the street. One of them shouted, "Hey, you want to fight him? You want to fight him now?"

"Sure, I'll fight him," I yelled. Thinking, *Now we'll see.* If it's one-on-one, I'm okay. I'm not afraid of going one-on-one, I've got no problem with that. But I didn't think it was going to be just the two of us. I was scared of being ganged up on by the whole crew over there.

By now the knot had grown into a little crowd. My friends took my bag and I squared off with the kid. Then, without thinking, I started transforming, just like the Decepticon guy back in the psych ward showed me. I'm moving my arms, I'm dipping, turning, moving, and I hear people saying, "What's he doing?" And right then I punch the kid, *boom*. He backs up, comes forward, and I grab him. But all of sudden I'm getting jumped from behind. I take the guy down to the ground, but they're kicking me in the head. Now I'm on top of this kid and everyone else is on top of me. I'm pounding his head on the ground, the adrenaline pumping so hard I don't feel anything of what's happening to me. I only know I have the guy I want down on the pavement with his face against the cement and I'm punching him. But at the same time I'm getting pummeled by a crowd of people on my back, and I hear my friends yelling, "You don't got to kick him in the head!" and I'm thinking, *Why are you yelling? Stop yelling. Do something!* And eventually they must have pulled them off me because I was on my feet again catching my breath and the crowd was already drifting off.

Afterward we were walking away and my friends were saying, "Oh, man, they were kicking you in the head. Those guys were stomping you." "Yeah?" I said, "so why didn't you stop them?" "Well, we tried to pull them off." I had been kicked pretty good, but the rush was so unbelievable I hardly knew it.

The next day back at school, the kid's face was all scraped up where I had had it against the ground. So in the end I was less bruised than he was. He had the marks. But still, I had been jumped by all his friends and family who lived in the neighborhood. I knew it wasn't going to be too safe for me around there anymore.

Soon after that I transferred out of there and back to the school in my old neighborhood, Junior High School 117. One Seventeen was notorious for poor academics and lots of violence, which was probably one reason Grandma had sent me off to Flatbush. It was right next to what we called LG, the Lafayette Gardens projects, one of the worst in Brooklyn. And since most of the kids in school were from LG there was a big gang effect. Luckily, when I went to 117 there were also a lot of kids I knew, so I didn't feel so marooned.

After I got there I found that 117 wasn't nearly as bad as it was made out to be. I knew people there, so I was hanging out with my friends, which made it comfortable. Plus, since the school had such a bad reputation most of the teachers were low seniority and tended not to be too experienced. Either that or they were near retirement and seemed more or less resigned to how things were. That made for a relaxed atmosphere where I felt I could be freer in the way I acted, I could be more myself. And that seemed to bring out the performer in me.

I think that an interest in acting was always down inside me somewhere. I hadn't done anything of that kind in school before, but I had always been eager to perform, like when I was little and would pretend I was Michael Jackson. I would dance Michael's dances and sing his songs, but what I was really concentrating on was being the character, on becoming the person I was imitating.

At 117 I especially enjoyed reading class. For this class we'd split up into groups of four or five according to reading level, with a couple of teachers and aides sitting in with the different groups. Sometimes it was made up of just our class, other times they'd put two classes together so there'd be a lot of kids. In general this class was looser, with more freedom, and since my reading had always been pretty good I was in one of the higher-level groups.

One day toward the end of class I was complaining to my group. I was saying how we, the students, were being treated unfairly. "You know," I was saying, "it's because they don't really want us to succeed. They don't want black people to succeed. They don't want Spanish people to succeed." Our teacher in that class was Italian-American, one of only three white teachers in the school. I'd gotten a little performance going for my group, and she thought it was amusing. After all, there she was, a white person, doing her best to help us read. So she encouraged me. "Go ahead, keep talking," she said, laughing. At first I was just talking to the group. But then I raised my voice a bit and stood up. When I did that, the whole class turned around, all the groups were looking at me.

"Look at a pool table, right?" I was saying. "The pool table represents the earth, green for the earth. You have all the different colored balls—the different ethnicities on the earth. Then you have the white ball. And the white ball's knocking all the other colors off the earth. And in the final play, what do you have? I said, *What do you have?* In the final play the white ball knocks the black ball off the table, off the earth. Now, why is this? What does this represent? Well, I have a dream. I have a dream that there is no pool table. I have a dream that there is no black ball. There is no white ball. I have a dream, yes I do. Yes I do." And people are going, "Yes! Yes! Preach on! Preach on!" I grabbed one of the colored milk crates. I stood up on it. "I have a *dream!*" I shouted, "I have a *dream*. Yes I *do!*" And I preached on that milk crate until the class ended.

After that it was not unusual to see me standing up in the middle of the class giving a speech about something I thought was important. A lot of it was humorous, but I was serious too. Before long I got the reputation of being a character. People had nicknames for me: I was "The Preacher," or "The Politician," or "Malcolm X Jr." I gave one famous I Have a Dream speech, ranting, "I have a dream of a school with no homework! I have a dream of students who can roam the halls free! Yes, yes, yes, I have a dream!"

When I was giving one of my I Have a Dream speeches I would modulate my voice. I'd seen clips of Martin Luther King Jr.'s I Have a Dream speech many times. I knew how to hit the key points, I had the rhythm down. And my voice had changed early. So it was deep, like a man's, like Martin's.

Because of my ability to speak and to animate my words the faculty at 117 chose me to go to Albany as part of a big statewide conference on improving student performance. This was a high-profile event with different educational groups, student speakers, and a panel of important politicians and educators.

When it came my turn to speak I went to the podium in front of this huge auditorium filled with people. The ten- or twelve-member panel was sitting up on stage. A number of speakers had gone before me, and the proceedings hadn't exactly been the most exciting. I stood there behind the microphone and I looked at the panel. One person was busy writing something, not even aware a new speaker was about to start. Another was talking to his neighbor. A third looked like he was asleep. So I said into the microphone, "Excuse me." They still weren't paying attention. "EXCUSE ME!" I said. "EXCUSE ME! NOW IS THE TIME TO LISTEN!"

By now they were all actually listening. That was all I needed to go off about how now is the time to listen to the students, time to listen to the people whose educations were at stake, to listen to what we think is best for our own education, for our own ability to learn. "You have to understand this," I said. "You have to respect what we are saying and what we think we need. You are sitting up here. Can you please stop talking and listen to me. Can you please put down your pen and listen to me. Because now is the time to listen!" When I finished, the whole auditorium was standing and clapping. And that was the moment when I knew that I had to be in the public eye. I just had to. I knew without any doubt that this is what I was meant to do. I was meant to perform.

Something had obviously happened to me. I might have had an acting tendency somewhere underneath, but most of my life I had been this quiet kid. Somehow, though, when I got to junior high school I felt freer. Maybe part of it was that now I had an audience, a group of people who would listen to what I had to say, and gave me a chance to say it.

Another part may have been that acting, or speaking, was a way for me to get out of myself. Because when I was speaking I wasn't me anymore. I was Malcolm X, or Martin Luther King Jr. I could speak about anything, because it wasn't really me doing it; it was a character. While I wouldn't normally talk about something like the pool table story, when I was in the preacher character it was a great story to tell. And while I might have been someone else delivering the words, it was Ejovi who was getting the attention and applause. After Albany I just knew that this is what I was made to do. And when a drama program started at school, with an instructor coming in from Pratt Institute, I joined up instantly.

My reading teacher was easy to talk to. Mrs. Anderson was interested in us and didn't get uptight if you kidded with her a little. My best friend at 117 was Allah Cunningham, and he and I sometimes stayed after class to talk to her. One day I found myself joking around with her, saying, "You're white, so you must be rich, right?" "Well," she said, "I have a house and a car." "Is it a big house?" I said. "How many cars? Can we come see your house sometime?" She had an apartment in New York City, but on weekends she'd go to her house in Albany. "Well," she said, "you know, maybe I will bring you boys up sometime."

After she said that we didn't let it rest. We kept bugging her about going up to her house and finally she said, "Okay. If your parents think it's all right, I'll bring you up there." So Grandma met her and thought about it some, then said, "Sure, don't see why not." And Allah's parents said the same, that it was fine with them too.

So we went to Albany the first of several times. Mrs. Anderson's neighborhood was great, all single houses. Hers wasn't a mansion, but it looked huge to us, two floors and a basement. Allah and I had always lived in cramped apartments, so that was new. We woke up in the morning and Mrs. Anderson cooked us a huge breakfast. I was used to rushing through a bowl of milk and cereal, so even that was a big change. There actually wasn't much to do in Albany, but that was okay too. Mrs. Anderson's son was in high school, a couple of years older than us. He drove us around, let us hang with him, and took us out to teen nights at the clubs.

Then there were the Albany characters; at least, to us they were characters. A big New York Giants football player lived next door. We only got to say hello to him, but just meeting someone that size was impressive. Another of her friends was a priest who wrote a series of books about a man named Joshua, a modern-day Jesus. We'd go out to the priest's place, a huge house on top of a hill with a big pond and a driveway that curved up to the house. In the living room of the priest's house there was an Italian guy playing the piano, with a gold pinky ring and a thick gold chain, straight out of a mafia movie. Another friend of Mrs. Anderson made her own jelly. Canned it and sold it. We got a bottle to take home for free. The whole thing was an experience. I thought it was great to see a different environment, a style of living that was so foreign to us. That visit and the ones that followed gave us a peek at the world beyond the one we knew in our part of Brooklyn.

■   ■   ■

At school there was a computer lab equipped with old IBM PCs. These machines had a couple of learning applications on them and a few games. You could use the typing tutors to improve your typing or play Wheel of Fortune, a word game where you'd guess words from selected letters. Most of us thought it was pretty cool to play around on the computers, but there wasn't that much to it and I wasn't all that enthusiastic. These were archaic machines that didn't allow you to do things that took any thought or creativity.

But the next year the principal and assistant principal got Apple computers in their offices as part of a city program that gave Macs to schools. By then the assistant principal, Mr. Forrest, was a friend of mine. I had been sent to his office a couple of times for doing something wrong, and he had befriended me. After we got to know each other he told me I was welcome in his office whenever I wanted to stop by.

Mr. Forrest had his own Apple at home, so he was already familiar with how this kind of computer worked. I would go into his office, maybe during lunchtime or if I had a free period, and hang out. He showed me how to use applications, how to make Word documents, how to make a flyer. He showed me how to run everything the Apple had on it. Sometimes if he had a meeting to go to he would just close the door and leave me alone to play on it.

In the computer lab, you would get there and the teacher would already have a program running that we would use. Because we were using PCs, if you wanted another program you had to exit, then type in the new program's name, then hit enter. But the Apple was completely different. It wasn't command-driven, like the PC. Instead it was what I later learned to call GUI (pronounced *gooey*) based. GUI—graphical user interface—meant that there were icons and buttons I could use, windows I could open and close, graphics I could click on and move around. Now I actually had control over what I would use and what I would do. The Apple gave me much more freedom to explore.

I did everything on that computer. I typed up my homework, my reports, little playscripts I might be thinking about. I experimented. There was a program called Quark XPress that allowed me to make flyers, posters, postcards, birthday cards. I always managed to find a reason to make a card or a flyer. Christmastime I'd make Christmas cards, at Easter time, Easter cards. You could print them so they'd fold into a card, with an inside and outside. I even made business cards for myself. The assistant principal had a business card program on his computer, and had made his own card: ALFONSO FORREST, VICE PRINCIPAL. Mine read EJOVI BARDEN (whenever I could, I liked to use my mother's maiden name), then gave my phone number. There was nothing else for me to put there. After a couple of months of this I had become the student computer expert, the one who was called on to show other kids how to do things when they needed help.

Knowing how to make flyers on the Apple got me thinking entrepreneurially. I'd had the experience of making a lot of money in my old supermarket bagging job, and now it occurred to me that maybe I could earn money making up advertising handbills for businesses in my area. A couple of blocks from my house a new Laundromat was just opening, so I approached the owner and proposed to make promotional flyers for him, and he gave me a chance. I made up a couple of test pieces. GRAND OPENING, they read, and beneath that, LAUNDROMAT, with a graphic of a washing machine in the middle, followed by COME SEE US. They weren't quite professional, but they weren't bad, either.

This was in 1993 and there weren't that many people doing graphics on computers, so for the time they were actually pretty good. In the end he didn't buy anything from me, but now I could see the possibility of making money with this thing. All you really needed was initiative and some creativity. Then you had to be willing to go out there, take the risk, and try out new things.

I didn't have an actual business yet, but I did want to get rich. Grandma was giving me a dollar a day to buy whatever I needed. That was more than the quarter she used to give me, but it wasn't exactly an allowance. So I needed money. Of course, a thirteen-year-old in my neighborhood could make money in drugs, but I didn't want to get involved in that business. I was revolted by it. Plus I thought I had too much going for me. At least that's what my teachers were telling me, and I believed them. The teachers and everyone else were saying that I was going to go somewhere, and in my opinion selling drugs wasn't going somewhere.

So, I thought, if you have a good idea, an original idea, you can make money. All I needed was that original idea.

6

Graphic Arts

The summer when I was fourteen I was walking through Clinton Hills, the neighborhood next to Bed Stuy, when I noticed big lights, trailers, cables, cameras, and a lot of people gathered around. It looked like they were shooting a movie. When I asked, they said it wasn't a movie, it was a TV show, *Ghostwriter*, that aired on channel 13, the New York City public station.

I knew the show, even if I didn't watch it that much. How cool would it be, I thought, to work on something like that? The whole thing was awesome. The cameras, the security, the staff people going around doing whatever they were doing, the dressing rooms. I saw the child actors, some of whom I recognized from the show. They even had their own chairs, and they were just kids, some of them younger than me.

Somehow I managed to slip past security. I talked with the kids, talked to some of the production assistants, some of the other staff people. I asked a lot of questions. What is it that you do, exactly? What could I do? How can I get on here? Finally one guy said, "All you need is to take a photograph and give it to us."

When he said that I ran home, dug out our old Polaroid, and had someone take a picture of me. Then I ran back and gave it to the guy along with my phone number. After that I started telling my friends, "Yeah, I'm going to the *Ghostwriter* television set." And all of a sudden it was, "Ejovi the actor! Ejovi the actor! The TV guy! Hey, when you gonna be on TV, Mr. TV guy?"

"Yeah," I said, "in a couple of weeks. Couple of weeks, I'm gonna be on *Ghostwriter*. They're going to air it." In my mind I could already see myself famous, like Michael Jackson, with millions of fans adoring me.

For two or three days I waited for them to call. When they didn't, I went back to Clinton Hills looking for them. A block or two from where they had been I found a set, but it wasn't them. When I asked, one of the people said they were shooting *Clockers*, Spike Lee's new movie. "Really?" I said. "How could I get some work?" "See that guy there?" he said, pointing to a stocky, light-skinned guy with a bullhorn on his shoulder and a big belt clip with two walkie-talkies. "Ask him. He's the one who does the hiring."

So I walked over to this guy, who I later found out was Yaya Lyons, the second assistant director. "How can I help out?" I said. "How can I be an actor? I'll do anything. Just tell me what and I'll do it." "Come back tomorrow," he said, "I'll see if I can find something for you."

At eight the next morning I was back, asking if there was anything for me to do now. "No, not now," he said. "Maybe later." But I was on the set, so I just stayed, all day, mainly next to the catering table. Eventually Yaya noticed me and said, "I want you to run over there and tell the extras to get ready because we're going to call them on."

All of a sudden, I'm working. I run down to the holding area and tell the production assistant who's watching the extras, "Okay, we're almost ready to go. Yaya told me to come down and tell you guys." I run back up to the set. It's dark already, almost ten o'clock. I have absolutely no idea where the time went.

"Yaya, need me to do anything else?"

"Yeah. Stand on that corner and make sure nobody comes past this way. When they yell 'Rolling,' make sure no one comes by. When you hear 'Cut,' you can let them come."

I hear someone yell, "Rolling!" then I recognize Spike Lee's voice saying, "Action!" I duck down. I'm making sure no one gets by. "Hold it there. Hold it, please." Then I hear, "Cut!" "Okay," I say, "you can go now." A minute later Spike walks by to talk to the actors. I'm standing there like a guard, feeling big and bad. I'm fourteen years old and I'm working for Spike Lee.

I can't believe it's gotten so late; I've been here all day long. I run off to call my grandmother from a pay phone. "Grandma, I'm here working on the set. It's going to go late, maybe until twelve. I'm going to need to be here."

She's yelling, "What? What set? What are you talking about? You get home right now!"

"Don't worry, Grandma, I'll be home soon. I'm on the set."

"You've got one hour!"

I left at eleven. When I asked Yaya where they were going to be the next day, he gave me a call sheet that listed all the locations and filming dates. "Be here in the morning," he told me, "as early as you can make it. We'll see what we can do."

First thing in the morning, I was there. And the day after and every day after that, until they closed the set down. I didn't get paid, but it didn't matter. I loved it. I was part of the team. I became friends with one of the stars, the child actor who played the drug dealer's little buddy. I met Delroy Lindo. I met Spike Lee. I got pictures of myself with them. I was on cloud nine.

I also met the security guys, a Black Muslim group called X-Men Security. There was one guy, short, but really tough looking, a mean look on his face. Not a guy you wanted to go near. It turned out that he was a martial arts specialist, a master in jujitsu, karate, and aikido, and the X-Men were all his students.

Near the end of the shoot I introduced myself to him. "I heard you have a school," I said. "You think I could come down and learn?"

"Yeah, sure," he said. "Come down and I'll show you a couple of things."

When I went to his place he was teaching aikido, which is all about leverage, taking someone else's aggression and using it for yourself. I watched while he demonstrated, choosing the biggest guy in the crowd and telling him, "Okay, grab me." Then he took the guy and flipped him, *slam*, flat on the ground. I'm thinking, *That's pretty cool. If I'm going to be able to fight, I'll have to know something like this.*

For the next month or so I trained in aikido with him, learning the basics. Right from the start it felt natural to me. The movements came easily, the blocks, the locks, the throws. When I was practicing these things I felt balanced and quick. I felt like I was in my element.

■ ■ ■

Aikido was good training for the High School of Graphic Communication Arts, at which I started that fall. One of the really aggressive martial arts styles would have been even better, because Graphic Arts was almost as much like a prison as a school. I had decided to go there because I'd loved to draw from the time I was a kid, and now I had also gotten some experience in computer graphics, which made me even more enthusiastic. At 117 I had experimented with setting up my own computer design business, and becoming an artist of some kind when I grew up sounded like a good idea. Of the things I really liked, art seemed like something I might be able to make a living at. That's why I thought Graphic Arts would be the right place. But I had no idea what it was actually going to be like in that school.

The High School of Graphic Arts was rough. It was a place where it paid to know how to fight. Like prisons, it was controlled by gangs, the main one being the Decepticons. For some reason Decepticon gang members had started going to Graphic Arts years ago and over time the school had become a magnet for them. Some of them might actually have been interested in graphic arts—there were a lot of graffiti artists in that school—but they mainly went because Graphic Arts was known as a Decepticon school.

Although the Decepticons were big where I came from, their real stronghold was Harlem and the Bronx, uptown, and that's where most of the Graphic Arts Decepticons were from. You'd see them in the hallways, uptown guys with the uptown look. The knit cap, not pulled down but worn more on the top of the head, tilted to the side. Big dark coats. Pants worn in a certain way, maybe cuffed at the bottom, sitting at just the right height on the hips. Blue was their favorite color, and they wore their gold chains in a particular fashion—all of which told you they were from uptown. Of course, after a while you just knew their faces.

The Decepticons traveled in groups and they were always picking fights. The most common sight at Graphic Arts was a bunch of Decepticons beating the crap out of some kid. In the hallways, in the lunchroom, in the bathrooms—places where you never, ever went by yourself. They weren't choosy; they picked on anybody who seemed like a good target. "What are you lookin' at?" "I'm not lookin' at anything." "Nah, nah, what are you lookin' at?" And all of a sudden you've got ten guys on top of someone mauling him. It was scary.

Like in jail, you had to watch your back. But after a while, also like jail, you knew there were certain people you just didn't look at. There were certain people who if they said something to you, you just kept walking. Even the security guards had their own stake in things. A lot of guards hung out with the Decepticons, smoking weed in the stairwells. At Graphic Arts you could hardly depend on security to protect you.

So it was natural that other groups were also organizing. The Latin Kings, the Nietas, the Zulu Nation—they were all there. You had to have someone to get down with. If you weren't part of a gang or a crew, you were going to be victimized. The Latin Kings were all Spanish, they emphasized Hispanic pride. The Nietas were Spanish and black, ethnic people, basically. The Zulu Nation included anyone who wanted to join—mostly black and Spanish, but white too. If you were a white guy who wanted to join a gang, Zulu was the only place to go. Being in a gang gave you a certain aura of prestige, a certain don't-mess-with-me air. It struck a little fear in people's hearts when they saw the look, the gang insignia, when they knew that you belonged to something big.

The uptown guys only associated with themselves, so everyone I knew started joining up mainly with the Latin Kings or the Nietas. A lot of people in my classes were Latin Kings and there was bad blood between them and the Decepticons. If two Latin Kings were walking along a hallway and ran into a pack of Decepticons, they would get beat up, and vice versa. There were larger eruptions too. For a week or two there might be almost a regular gang war. The school had metal detectors, so there were no guns. But there were plenty of razors, box cutters, and knives, which were easier to sneak in. You had to watch yourself.

I had never been in a gang, except for our Autobots when I was a kid. But now it was more serious. After a while I thought that just as a matter of self-preservation I'd better get in with somebody. So along with a couple of my friends I joined the Zulu Nation.

The way you got into Zulu was different from how you got into the other gangs. The others jumped you in. If a guy wanted to join the Decepticons, they'd tell him that if he wanted to be down with them he had to go through the initiation. Once in a while you'd see these initiations happening in a bathroom or back corridor. The guy who wanted in would be in the middle of a circle. He'd throw up his hands and say, "All right, I'm ready. I want to be down." Then the biggest guy would step into the circle with him. And the biggest guy was the toughest guy, maybe the leader, the badass guy. The guy who wants to join would go to wale on the big guy, and the big guy would go *boom, boom*, knock him down. All of a sudden the fifteen guys who were standing around would be stomping him in the head, kicking him in the back, punching him. Then, after five or ten minutes of beating the crap out of him, they'd pick him up, give him a hug, and tell him he's in, he's part of the gang. That's how you got into the Decepticons, and it was the same for some of the others.

But Zulu Nation didn't jump you in. Instead, I signed a piece of paper and received "The Lessons," twenty pages or so of rules and philosophy. These are the lessons, they told me, this is the knowledge you must study if you want to part of the Zulu Nation. Learn them. Be ready to talk about them at the next meeting.

I read them. You must be kind to your brothers and sisters, they said. You must treat all people with respect, no matter who they are. You must respect yourself, you must respect your family. They were lessons about basic morality. Lessons about karma. If you do such and such to people you should expect the same thing to come around to you. Lessons about looking out for each other.

Zulu Nation, like the others, was more than a gang, it was an organization whose main purpose was unity. A lot of people who join gangs just don't have family, or their family situation is really bad. The gangs give them something to belong to. They give them protection, they give them love. They might give them a girlfriend. You could find your ideal mate in the gang. And when one member of a group got romantically involved with another member there was a special ceremony, almost as if you were promoted to a higher rank. Relationships like that highlighted the family or clan aspect of the organization.

Each of the gangs had its own identity. If the Latin Kings were about Hispanic pride and Nietas was for poor ethnic people more generally, Zulu Nation was most famous for its hip-hop origin and associations. Afrika Bambaattaa, the overall head of Zulu, was one of hip-hop's founding fathers, an artist whose music and years of performances helped make hip-hop a cultural phenomenon. But the organization was broken down into chapters, with anywhere between fifty and three hundred members per chapter, and each different type of chapter had its own specialty.

Some were specifically hip-hop chapters. Members of these chapters were artists and DJs and break-dancers—over the years Zulu Nation has had many famous break-dancers. They practiced, went dancing, performed, did music promotion at different events. They spread the culture.

Then there were the knowledge chapters, whose agenda was to discuss and spread information about economic and social ideas. The basic concepts on which they concentrated were mainly how to improve society and better your family, how to become a better person. But they also talked a lot about conspiracy theories. The Illuminati, secret controllers of the world, were a big subject along with alien technology, embedded microchips and bar codes, the coming of the New World Order, and 666, the sign of the beast. I would sometimes go to a knowledge chapter in Brooklyn that only dealt with New Age and technology. They'd have newspaper articles about the paperless society, about how microchips are being used to keep track of pets and how they might be used in people.

The emphasis on conspiracy theories might sound paranoid, but it spreads from the idea that things are basically controlled by The Man. That doesn't seem paranoid at all to people from poor neighborhoods who have felt oppressed for generations. From their point of view there already is a caste system in place and technology is coming on the scene that will make the caste system permanent, create mechanisms for tracking people and categorizing them, mechanisms that are controlled by secret knowledge and unseen hands. So if you're poor, for instance, and your ID says you're poor, there will be no way to make it out of that caste. They talked about a future that in some ways has actually materialized. Today, for example, people on welfare don't use food stamps, they don't get welfare checks; instead, they are given a little card, like a credit card. You go into a store and use it like a credit card. You swipe the card and enter your PIN and if you're okay with the welfare system, the transaction is approved. But if the system says you're cut off, then that's it, you don't eat. The government controls what you can eat. With technology like that it's an easy step to wondering about how else it might be used and who might use it, and to what purposes.

Sometimes I went to the knowledge chapter meetings, because I was interested in technology and I definitely had my own touch of paranoia going back to the times when I had to be a lookout for my mother. But I wasn't in a knowledge chapter, I was in the Elites, a special chapter dedicated to fighting. I got into the Elites more or less by accident. There were Elites at my school and one of my friends was in them, so that's who I fell in with.

The Elites were special. The Zulu Nation symbol was a leather patch in the shape of Africa worn on a string around the neck. For most Zulus the patch was red, green, and black, African colors. But Elite patches were all black. We had our own chapters. We were known, but we weren't talked about. When we came on the scene all of a sudden it would be like, "Oh." People acted deferential. If you had your black patch on they would look at you while trying to seem like they weren't looking. We had everyone's respect.

Our job was one thing—fighting. Basically, we were the police of the organization, the enforcers. When a chapter in Brooklyn had a problem, maybe they were being threatened by a local gang or crew, we were the ones who went over there and did the fighting for them.

The Elites were tough guys. We had regular meetings, once a month, sometimes more, when all the Elites would get together, two or three hundred strong, at a park in the Bronx. Everyone would gather in a circle to hear the head of the Elites talk about what we needed to do during the next few weeks, where we were going to be going, where we were going to be fighting.

When he was finished with the schedule of events, the leader would say, "Who standing in this circle considers himself to be a good fighter? Who? Tell me right now! I want to know, who's the best fighter in this circle?"

Twenty guys would raise their hand. "I'm the best!" The leader would walk around, saying, "Okay, you. Come into the middle of the circle. You the best fighter? Come into the middle of the circle."

The guy would come into the middle.

The leader: "Look at this man! This is the bravest man in this circle! Why can't everyone else here be like this man? You have to be ready to fight at any time. You have to be ready to throw down. Anytime! Just like this man here. You understand what it means to be part of the Elites? You must be willing to fight!" Then, to the guy who had stepped forward, "Are you ready to throw down?"

"Yeah, anytime."

"Okay." The leader walks around the circle, says, "You, you,

you, you, you, and you. Go fuck him up!" And they would go beat the daylights out of the guy.

That happened every meeting. But after the guy got up off the ground, he got a lot of respect. He knew what was coming, and he really *was* a tough guy. He knew what he was in for. We knew that we were there to fight.

Throughout the city there was more or less a truce among the Nietas, the Latin Kings, and Zulu Nation. The idea was that if I was in a situation with a Decepticon in Harlem, in the Bronx, or anywhere, a Latin King or a Nieta would have to help me because there was a truce and we were all brothers. Of course, it didn't always work that way, because certain chapters had problems with other chapters. Maybe one chapter of Zulu would be having a beef with a chapter of Latin Kings. And all of a sudden if you go into that chapter's area, you're in danger. So it was a little risky at times, because you might not know who was respecting the truce and who wasn't.

But mostly it worked. We had our own Zulu handshake, you give it and touch your heart. "Peace," you say, "*aquí.*" But there was also the special universal handshake that would identify you as a member of the truce. A Latin King might say, "How do I know you're Zulu? What's the universal handshake?" And I'd do the handshake with him that you had to know if you were a member of any of these groups. Shake, shake, thumb, around the pinkie, then back over the thumb. Hand on heart, index finger and pinkie extended. A Latin King would do the same, but he'd use his thumb too. The three fingers, thumb, index, pinkie are the crown, the crown of the Latin Kings, across your heart. Their symbol across their heart, your symbol across yours. Brotherhood.

■ ■ ■

While I was seeing the inside of the gang world, I was also finding out about oppression and brotherhood from another perspective, a political one.

My first year at Graphic Arts, 1995, was a year of drastic cuts in the New York school budget. A lot of people were upset about it, but one of my teachers was really agitated. He and I got along well and one day he gave me a flyer announcing that there was going to be a protest against the cuts down at City Hall. "This is about students' rights," he told me. "You should go."

Well, if he was talking about students' rights, he was talking to the right person; I mean, I was all about students' rights before this. Back at 117 I had spent half my time standing on a milk crate. I had given that speech in Albany in front of those experts. And back in junior high an incident happened that really radicalized me.

One of our teachers there was known to have a bad temper. She had hit students before, and one day she smacked me. I can't remember what it was for, I must have been doing something. But it couldn't have been that bad, and when she hit me I went wild. Not physically, but emotionally. I was dumbfounded. I complained to everybody in the school. Loudly. "How in the world can a teacher smack a student?!" I said this to the principal. "How can a teacher smack a student and keep her job?"

What pissed me off even more was the response I got from the principal and assistant principal, both of whom I considered my friends. They turned it around and made it my fault. "You should have been sitting down," they said. "That's just the way Mrs. So-

and-So is; what were *you* doing?" Then they went to speak to my grandmother at home. I'm telling my grandmother, "The teacher did this!" I'm telling the principal and assistant principal, "I want you guys to apologize! I want the teacher to apologize! I can't believe you're allowing this!" And they're saying to my grandmother, "Aside from the teacher issue, we're concerned about whether everything in Ejovi's home environment is okay." And I'm thinking to myself, *In my house?* I can't believe you're saying this. What do you mean "home environment"? You're going to come to my house, after a teacher smacks me, and you're going to question my grandmother about my home environment? You've got to be nuts!

After that I was really attuned to how students were treated. So when my teacher at Graphic Arts gave me this flyer about the protest, I was all for it. I was upset about the way students were being treated anyway, and now the city was slashing the educational budget. How could they do that? I hopped right on the bandwagon, and along with another student I became one of the main organizers for the rally.

Our big effort was to organize a cut day, a day when the whole student body would skip school to go down to City Hall and demonstrate. This teacher, who was a real radical, was our main supporter. I made up flyers in the computer lab and gave them to him. He took them home, maybe improved the format a little, then made hundreds of copies. I put the flyers up at school, I handed them out, I stuck them up all over the stations on the subway lines most students took, the C and the A, which went through three boroughs. I talked to people, I spread the word. I had friends in different high schools and I told them, "Look, tell your friends that everyone in the city's cutting. Tell them to get down to City Hall for the demonstration." I figured that maybe we could get four or five schools to walk out.

A *Daily News* reporter picked up on what was happening, and interviewed me. She talked to me while I was organizing it and wrote:

> *Public high school students, seething over huge budget cuts, plan to walk out of school Thursday and march in protest to City Hall. News of the walkout has spread like wildfire since organizers at Graphic Arts High School in Manhattan began getting the word out.*

Judging from the responses I was getting and all the enthusiasm of those around me, I thought we'd have a pretty big turnout, but when I got down to City Hall Plaza that Thursday I was amazed. There were so many people you couldn't even walk. I had thought that we were the main ones organizing this, but here was this gigantic mass of protestors, including college students, some older people, speakers with bullhorns. It was exciting, but a little deflating too. I thought, *You know, there must have been people other than us organizing this thing.* I didn't feel quite as important as I had, though I still thought I must have made a difference.

At some point during the demonstration the police decided it was time to clear out the crowd. But not everybody wanted to move peacefully, and when the cops started getting resistance they began Macing people and wading into them with nightsticks. Mounted police charged into the crowds and people started running for dear life. In a minute the protest turned into a riot. The movement of the crowd picked me up off my feet and rushed me along, the crush of people screaming and running. Mace hung heavy in the air. People around me were getting facefuls of it, which I managed to avoid, but I was getting powerful whiffs and my eyes were tearing up. Everyone was pushing and screaming, trying to get away from this, and the cops were yelling, "Get back! Get back! Break it up, break it up!" But they closed up the train station so there was nowhere to go except the immediate area. On every side of me people were being assaulted by the police, dragged around and thrown inside riot trucks.

That night the *Daily News* writer called me at home and asked me what happened. I told her my story, and next day the paper reported how the police had Maced and abused students at City Hall:

> *Cops in riot gear clashed with the more than 8,000 angry high school and college students who descended on City Hall yesterday to protest budget cuts. Ejovi Barden, a student at Graphic Arts High School in Manhattan, said police attacked him with tear gas and that he got caught in a crowd of pushing protestors as they were being herded through barricades.*

After that the teacher who had helped us put this together said, "You know, maybe you should come to some of our meetings." I went, partly because I was curious, but also because I had met his daughter, who was my age and very attractive. The meeting was in a factory building on Twenty-seventh Street, right across from the Fashion Institute of Technology. Inside the building there was a newspaper office; THE DAILY WORKER, the sign read. That's when I started realizing, hey, this is the Communist Party. So that's how I found myself sitting in the offices of the Party newspaper discussing the sad condition of working people in America.

I had never known anything about communism before, but if you took what Malcolm X and Martin Luther King Jr. had said about how we were being oppressed, it seemed like basically the same theory. Except now it wasn't the white man oppressing the black man, it was the capitalists oppressing the workers. The same general theory, except more universal. It didn't sound foreign to me at all.

I was introduced to a lot of things I didn't know about, such as the fact that there are thousands and thousands of illegal immigrants in Manhattan working in sweatshops under terrible conditions. They don't speak English; they don't know anything about their rights. And when you try to organize them, you can't get anywhere. People who were distributing the newspaper talked about how difficult it was to get the sweatshop workers to even pick up the paper, because they were so afraid of losing their jobs. Or getting deported.

People at the meetings talked about these kinds of things. They said, "Look at why it is that so many minorities can't make it out of the neighborhoods. It's not because the white man is oppressing them. It's because the way the system is set up, it prevents people who aren't already middle class or above from making it." We were in the newspaper offices, and the stories were coming from all over the country, all over the world. Stories of oppression. It was the same the world over, it seemed, no matter what color you were. No matter what country you were in. It was the plight of the working class. I identified with that. It appealed to my idealism and my sense of outrage. I was all for anything that would fight this universal injustice. We needed to get rid of the government, have the system run by the people, with a council of members run by the people. We needed to ensure students' rights, poor people's rights, oppressed people's rights. We needed to distribute the wealth evenly.

For a few months I was really excited by all this. But after a while the "distribute the wealth evenly" part got me to thinking. I had always wanted to be a business person, an entrepreneur, even if "always" only meant four or five years back, when I had earned all that money as a grocery bagger. Certainly I always wanted to make money. So I started thinking, sure, it would be great if we could distribute the wealth fairly and everyone could be equal. But look at it the other way. If I'm working eighteen-hour days to start up my business, I want to be paid according to the work I do. If someone is running a machine for eight hours a day, pay him accordingly. But if I'm working on my own for endless hours, getting no sleep, struggling to fulfill a dream that I created myself, well, I want to be paid accordingly, too. But in the ideal system my wealth would be evenly distributed, even to people who didn't work at all. If my ability was great, but someone else's needs were greater, he'd end up with the money. That didn't seem right at all. I couldn't grasp it.

So at one of the meetings I asked the teacher, "Wait. What about me? What if I decide to start my own business and I become rich? What's going to happen to me?"

"Well," he said, "you know, we can't allow people to be rich, because we need to distribute the wealth among all the people. You're going to have to give away your money."

"Uh, okay," I said. The part about rights said something to my sense of justice, as did the parts about fighting oppression and the need for altruism; those made a strong impression on me. I knew very well that poor people needed help and I even started thinking about how I might do that if I ever made enough money. But getting my wealth distributed evenly? That was something else. Others might think that was a good idea, but it wasn't for me. And after that meeting I never went back.

7

Administrator Jim

Mark Howard was one of the kids I had met on the *Clockers* set the summer before I started high school. He played a young drug dealer, a member of the star's crew. Mark and I hit it off and by the time September came around we were already best friends. We didn't go to the same school, but we lived near each other and hung out as much as we could. Mark even saved my life once.

One of our favorite activities was aggressive in-line skating—tricks, rails, stuff like that. One day we were skating in Clinton Hills on Lafayette Street and we grabbed onto the back of a bus. Even on city streets, a bus could take you flying. Catching a bus ride was one of the more exciting things you could do on skates.

I was wearing shorts that day, and when we grabbed on I didn't notice I had gotten myself positioned directly behind the exhaust pipe. Of course I realized it after the bus got going, but by then it was too late to move, and a minute later the exhaust started burning my legs so badly that I lost control of the skates and my legs went out from under me. I was holding on by one hand, yelling to Mark, "Help me, help me! Pick me up! Pick me up!" Of course, the bus driver had no idea we were there and he kept accelerating. I knew that if I fell I'd probably get run over—there's lots of traffic on the street. At the very least I was going to break something. We were up to about twenty-five miles an hour and still picking up speed. I was dragging my skates and barely holding on. I knew I couldn't last much longer. But just before my grip gave out Mark managed to grab my arm and pull me up, holding me with one hand while he held on himself with his other hand. He kept us both hanging there until the bus slowed down for the next stop and we were able to drop off. So that was my best friend, Mark.

By the time I finished working on the *Clockers* set I knew a lot more about movies than I had at first. I knew enough to get head shots and send them out to people, and I started getting a couple of little parts. One bit part I got was on *New York Undercover*, where I played a member of a gang headed by the bad guy, played by Sean Nelson. I was only on-screen for a few seconds, but afterward my phone was ringing off the hook with family and friends who had seen me. Sean was already well known for *Fresh*, a movie in which he had played the lead, *The Wood*, a show about drug addicts in Boston, and especially for his role in *The Corner*, a prizewinning HBO docudrama set in Baltimore's ghetto.

Sean went to the High School for Performing Arts, which was just three blocks away from Graphic Arts, down on Forty-sixth Street. Peewee Love, another *Clockers* actor I had gotten to know, was there, too. He had played the drug dealer's young friend. With Performing Arts so close, most days after class I used to go over and hang with Sean, Peewee, and some of the other young actors I had met. And the more I hung out there, the more I thought that this was the place where I really wanted to be. So when auditions were called for admission to the school, I went to try out, and I persuaded Mark to audition with me.

I had done theater in junior high school, so I basically knew how a script works and what an audition was, but I had never actually read for anything. In auditions it's common for actors to choose a part from a play and read, or recite. But I decided I'd write my own little monologue. It was something about a character from my neighborhood, a street thug. I'd seen enough of them in my life, so I thought I could play one well enough. "Yo, see what I'm sayin'? It's like, you walkin' down the street and some guy's grillin' you and, *bam*, you just want to take him out."— something along those lines. But while I was doing it, I was listening to myself and thinking how really bad this was. So in the middle of it I stopped and said to the person running the audition, "You know, I'm sorry I messed up a bit, but my agent just told me about these auditions and I'm not really prepared. I just came from another one. So it's really been hectic." The guy asked, "Oh, you have an agent? Who's your agent?"

"Uh, the Carson Organization," I said. I had heard the name somewhere and it just seemed to pop into my head.

"Right," he said. "Yes. The Carson Organization. I know them. Okay. Okay, okay. You sounded good. Great. Good. Okay. You're in."

So there I was. Of course I didn't have a real agent any more than I had a real monologue. It was just a bit of social engineering. The whole thing was an improvisation. But I was in. And so was Mark, who had also written something up for himself. Okay, I thought, now I'm starting a real acting career. I had no idea that what I was really starting was my hacking career.

At that time my uncle Osie was in graduate school at Pace University, but he was living at home. I'd gotten a pretty good introduction to computers from working on Mr. Forrest's Apple in junior high, and Osie was letting me use his computer and also his America Online account. So I knew how to get online and I was starting to find my way around certain parts of the internet. I had also read a little about hacking and I was curious about it.

One day at school I was talking to this kid who also had AOL. He was Indian, but he didn't talk like he was from India, he talked like a slacker skateboarder, which is what he was, with orange hair, his hat on backward, his slacker clothes, and his skateboard under his arm. "Oh wow," he said. "This is cool. This is rad. This is really neat, dude. Slammin'. You've got to check this out." Normally I never would have associated with this guy. This was not my type of person. This wasn't even my language.

His name was Anand, but he called himself 9X, like the phone company NYNEX. "Dude," he says, "you know warez trading?"

"Oh yeah," I said. "Warez. I know warez."

"Right," he says. "Get yourself some really cool software."

Now, I didn't know warez at all, though I figured it was something to do with software—ware, wares—warez, though he pronounced it like *where-ez*. Which is what it turned out to be, free software you could get by trading on the internet. Free, as in pirated.

Trading on the internet was something I already did know about, because what I had mostly been doing on Osie's computer was trading pictures—of naked girls. I had been in an AOL chatroom one day, just the standard chat: "Hi, I'm Jovi, from Brooklyn." "Hi, Jovi. I'm Denise from Akron, Ohio," when somebody in the chatroom posted, "If you want private pics, quick, come to my private room. XXX Pics."

So of course I went, and right away I found out that there were private porn pic chatrooms, or trading rooms. Someone was promoting his "XXX Pics." But over there was somebody else promoting his private room. "XXXX Pics. Better than XXX Pics." In these rooms you could ask for pictures of tall blonde women, for example, and twenty people would send pics of tall blondes to your e-mail address. Ask for busty brunettes, and you'd get twenty busty brunettes. And they'd all ask for something in return.

"This is Bob. I have Madonna, full frontal. You want her?"

"Yes."

"Okay, what do you have?"

"I have Brooke Shields."

"Okay, send me that."

Then I trade Bob's pic to you and yours to Bob. I ask, "Do you have anything else to send me?"

"Yeah, yeah, yeah. I like the pic you sent. I'll send you something else."

"Okay." I go back to Bob. "I like your pic. Anything else you can send me, Bob?"

"Yeah, I like yours. Send me something else."

So I send you Bob's and Bob yours, yours to Mack and Mack's to you, Bob's to Mack and Mack's to Bob. It goes on and on. Before you know it, you've got a hundred megs of adult pictures stored in your e-mail account.

So all this I knew about, e-mail and private rooms and trading, and now my new friend 9X is telling me, "Yeah, you can join this warez room and get all these cool programs."

So 9X gave me the name of a private chatroom—you couldn't just get there by yourself, you had to know the secret name from someone—and I went online, typed in the name, and there I was, inside. The first time I actually did this, it was amazing. There seemed to be about thirty people in the room, though it was so unbelievably hectic it was hard to tell. I'm there, but not saying anything, just looking. And everyone's offering software.

Dark Night: "I've got this and this and this."

    Job: "I've got this, this, this, this, and this. You want some of it?"

Zeroday: "I've got this cool program. If anyone wants it I'm willing to trade for such and such."

It's like the trading floor of the stock exchange. Everyone's trying to trade something. Everyone's trying to sell something to someone else. All of them at the same time. And I'm sitting there watching, dumbfounded. Totally mesmerized. Trying to get used to the flow so I can follow something. From the next room, Osie's yelling, "Get off the computer already!" And I'm like, "Yeah, okay." But I can't tear my eyes away. Outside the dealers are slanging, the gangsters are banging, but I'm oblivious. It's like the thing is alive. I could sit there and stare at it forever.

Before too long I was picking up the threads of who was doing what, and once I did I could see the obvious—namely, that porn trading and warez trading were essentially the same thing. You could trade adult pictures, or you could trade programs. All you're doing in either case is transferring files back and forth. It's all the same theory. You have something on your hard drive, pictures, pirated programs, there's no real difference. It's just that instead of "Madonna, full frontal" someone's going, "Photoshop. Photoshop."

As soon as I felt comfortable with how people in the warez room were doing things, I jumped in: "I've got Microsoft Word 4 [Osie's word processor]. If anyone wants to trade, send me a message with what you've got." A moment later, I've got a message: "Hey, I've got WordPerfect, it's better than Microsoft Word. Want to trade?" So I send e-mails to him with Microsoft Word attached and I hear *bing!* "You've got mail." I look and there's ten e-mails there—one file, but broken down into segments.

So now I've got two programs. I go back in. "I've got Microsoft Word and WordPerfect. If anyone wants to trade a graphics program, send me a message." And a moment later I'm reading something like: "I've got Adobe Graphics, Word Graphics, and Graphics Effects. I'm looking for WordPerfect. Can you send that?"

After that there was no stopping. I was on the computer every minute Osie wasn't. By the end of the day I might have three hundred e-mails waiting, all that software just sitting there. I was especially into graphic design programs, because I wanted to do web pages and design. I had never forgotten the idea of starting a design business, except that now I had professional software for it instead of a couple of prepackaged graphics programs. And these programs I was getting each cost two or three or four hundred dollars at a store. They still do. But in the warez rooms, all they cost was a trade. Osie would tell me, "I'm going to buy the new version of Microsoft Word." I'd say, "Oh, no, no. I can get a copy of that." Or the new version of Windows. "No, I can get that."

Of course I also began to think, Hey, maybe this isn't safe. Pictures are one thing, no one cares about pictures. But it was obvious that the programs I was getting were pirated. They cost money, they weren't supposed to be given away. Maybe it wasn't smart to be doing this on Osie's computer. But then again, it wasn't hard to rationalize either. I mean, there I was, this poor kid living in Bed Stuy. How else am I supposed to get these things?

Every day I'm going back to school and talking to this kid 9X. "Oh wow, this is so cool. I've got this new program." "Really? I want that program. Can you send it to me?" We're trading information. I'm telling him what I did the night before, he's telling me what he did. And one day he says, "You know about phishes, right?"

"Fishes?"

"Yeah, no. Phishes. Phishing. You know how to go phishing?"

"No, what's that?"

"Oh, man, it's the neatest. Listen, dude, this is what you do. You create a message and send it in Instant Messenger to all the people in a chatroom, and ask them for their credit cards. And they send you their cards. It's pretty cool."

"Really? They send you their cards?"

"Yeah. Then you take the credit cards, and you can buy clothes. You can do carding."

"What's carding?"

"Carding's like when you take the credit card and buy something online and have it shipped someplace where you can pick it up. So you don't have to get it at your normal house."

"That's cool."

"Yeah, yeah, dude. You should try it. Then you can take the cards and use a credit card generator."

"A what?"

"A credit card generator, man. You use a credit card generator to create a bunch of credit card numbers from one credit card that you get. A bunch of fake credit cards. And you can use those credit cards to create new AOL screen names. So you can keep regenerating, regenerating, and regenerating.

"You can?"

"Yeah, it's cool. You should try it."

Hmm, I was thinking, *is all this really possible?* With 9X showing me the way, I started experimenting. Back then, in 1995, before computer security began catching up, you could use words like "Admin" and "Staff." Here's what you'd do: First you create an account called "Administrator Jim." That's your AOL screen name. Then you send a message to someone in a chatroom from AdminJim. "Hi. This is Jim Reilly from AOL staff. We need to update our credit card records. Yours have been corrupted in our database. In order to continue uninterrupted service for your account, please send us your current credit card number, full name, and billing address."

Meanwhile, the user is in a chatroom having a little conversation about his day at work and *bing*, he gets a message from AdminJim. *Uh oh,* he thinks, *this must be important. This guy's on AOL staff. My service might be interrupted?* He replies, "Okay, here's my name: Bob Smith. Here's my credit card number, here's my billing address. I really hope you guys can get the system up and running, and thanks for letting me know."

That was the basic idea. What made it feasible was that you

could send this message to everyone in a chatroom at the same time, a mass message. Thirty messages from AdminJim asking everybody for their credit cards.

I was already familiar with the concept of mass messaging from the warez trading rooms. You'd be in there and suddenly you'd get a message: "If anyone wants the new version of such and such, press 999." And five minutes after I entered 999 the new version was in my in box.

These mass messages were generated by automated programs called "bots." A person who wanted to mass message a room would run a bot and every thirty seconds the bot would send, "If you want a free version . . ." or whatever else you might want to say to everyone who was logged on.

As I found out later, these bots were created mainly by young hackers or traders who wanted to make it easier to trade. Among the items available in the warez chatrooms was programming software like Visual Basic or C++. These were thousand-dollar compiling programs that Microsoft and the other producers sell to developers and programmers. You couldn't just duck into your corner computer shop for them. But the warez traders, mostly kids, were getting them for free. And once they had them they said to themselves, *Gee, I've got this thing, I might as well learn how to use it.*

So they went and bought a book about C++ for fifteen dollars, or maybe they got a book online from the warez guys. Or maybe they just had the manual that came with it, or the tutorial. So they read this over and they say, *Let's see, I want to create a window. So how do I do that?*

The program tutorial reads: "To create a window, enter the following code." It's a simple code that you always use to create a window. *Okay*, you say, *let me enter that code.* You do. Boom. Hey. *Okay, I've got a window. Cool. Now I want the window to have text in it.* And the tutorial tells you, "If you want to put text inside this window, add the following code and enter your text." Okay, so you need this code to create the window and that code to put text in the window.

So, you've got the window, you've got the text. Now, how do you actually send it to people? Here you've got lots of options. Maybe someone in the chatroom already knows how to do it, some twenty-year-old who writes code for a living. You ask and he'll tell you. He's done it a thousand times. Or maybe the tutorial will explain it, or you can go buy a book. But mainly you do lots and lots of trial and error. You explore. You find out you can create a button that reads "click OK" to go along with your message window, and another button that reads "click cancel."

And now you want to do something a little more advanced. You want to figure out how to send e-mail to everyone who enters a room. You already know the basic stuff. How to create a window, how to create text, how to create an OK button, a cancel button, how to send it to someone. Now you learn how to send a mass e-mail. You build up your skills in steps until you learn how to send the message to everyone in a room at the same time, and how to do it not manually, but automatically. So it sends itself. Now suddenly—except it took a ton of work to get here—you're a programmer. And you're still only fourteen years old. This is 1994, 1995 and there is no frenzy yet over internet IPOs and stock options. But you're a rudimentary programmer and you've got a whole universe out there that's just waiting for you to conquer it.

But the immediate object right now is to get credit cards. A chatroom might have twenty or thirty people in it. Running your bot you get to all of them, and you get to everyone who enters the room later. Since it's automatic, it doesn't take any time, which means you can go from one chatroom to the next. You go through five chatrooms thirty people deep and you send the mass message to everyone in each chatroom, one after the other. You keep doing it and doing it and next thing you know you're getting thirty, fifty, a hundred responses back with people's credit cards. A few people are saying, "I'm going to report you." But most respond. At the beginning almost *all* of them responded.

Of course, I wasn't sure what to do with the cards, except set up AOL accounts so I could be online all the time. But basically I was in hog heaven, almost in a kind of trance. Osie had no idea what I was doing, he just thought I was spending far too much time on his computer. Maybe I was addicted or something. But I hadn't been charging his account for a long time, so that wasn't bothering him. As long as he wasn't using his computer, I was welcome to.

By now I'd gotten over the porn stuff, I'd gotten over the warez stuff, and I was thinking, *I can be really good if I can just get this new stuff down.* I had free software, naked pictures galore. And now I had credit cards. Every day I'd talk to 9X and get ideas. I couldn't wait to get home, hop online, and start exploring. And what I wanted to see about now was credit card generators. I had heard people mentioning them in the warez rooms, where they were items of trade, and now 9X was telling me to get one. But I never knew what they were. Now that I was into phishing, it was time to find out.

8

What's Better Than Free?

Even the name "credit card generator" got me excited. By this time I was researching everything I heard about. It was all so completely amazing. So I went to a search engine on the Net, typed in "credit card generator," and found a couple of websites where there was some good, in-depth documentation on credit cards and generators.

I was reading the documentation and my eyes were open so wide, they were like saucers. It seems that the numbers on each credit card identify it in a particular way. One sequence of numbers signifies the type of card, MasterCard, Visa, whatever. Another identifies the issuing bank or credit card company, Citicorp, MBNA, Capital One. Another is the customer account number. Each card company then has its own unique formula, or algorithm, that takes these three groups of numbers and performs a set of calculations on them, which generates an additional sequence of numbers. The last number in this additional set is a digit called a "checksum." This checksum is used to make sure the entire sequence fits the card number profile of the issuing bank.

What's obvious after you learn this is that if you get hold of enough Capital One card numbers, you can crack the code; you can figure out the algorithm. A credit card generator is a computer program written by someone who knows what the identifying sequences are for, say, a Capital One Visa card, and who has also cracked the Capital One algorithm. Anyone who uses that program can generate thousands of apparently valid credit card numbers.

Of course, the cards aren't really valid. But anyone who checks the number will find that it fits the Capital One profile. Okay, now here's what happens when a store checks a credit card: Computers use binary code, sequences of numerical ones and zeros, to perform every single operation. So for every question someone asks a computer program, such as a credit card verification program, there's a numerical response, 1 or 0. In simplified terms, it goes something like this: The verification program asks, "Is this card whose number I'm scanning a valid card?" The reply is 1 for yes, 0 for no. The program checks card type number sequence—yes; bank identifying sequence—yes; algorithmic sequence—yes. Okay, next question: Can I debit fifty dollars from this account?—yes. Okay, next question: Is this the correct name for this account number?—yes. And so on. Lots of cross-checks. If everything comes back positive, the purchase is approved; if not, it's denied.

But in the mid-nineties, some online companies only ran the first verifying step: Do the identifying number sequences and checksums match the type of card, issuing bank, and that bank's proprietary algorithm? If that worked, you passed the test. They'd worry about billing you later. Back then a lot of internet sites, AOL being one of them, worked that way.

Once I understood what a generator did, I went into a warez room and traded for one. Then I entered a real credit card number I had phished into it and told the generator how many card numbers I wanted. The next thing I knew I had a hundred fake card numbers.

It didn't take long before I got to be an expert at credit cards. When you get twenty or thirty cards a day and analyze each of them, you learn fast. I was analyzing every credit card I got, just like I was analyzing every program I got. It was all so fascinating, a new world. I got to the point where I could identify most credit cards by looking at the first sequence of numbers. Oh, this one's Citibank, Twenty-first Street branch. This one's Chemical Bank, this one's Dime. Which is really cool information to have. You might never use it for anything, but it's knowledge, interesting trivia knowledge. Some guys might know the batting average of every New York Yankee. I knew the identification numbers of every credit card company. The weird thing was that I hardly knew anything else about credit cards. I didn't have one myself, of course. Other than Osie, I probably didn't even *know* anyone who had one.

What I mainly did with the cards was create AOL accounts. That way I could keep all the warez programs in e-mail accounts, using AOL's mail server for storage instead of downloading them to Osie's machine, which only had 400 megs of hard drive space anyway. I'd go through the entire application process, putting in all fake information. Name: Count Dracula. Address: 12 Freddyscomingforyou Street, Oregon, Michigan, 01234. I'd enter silly things, then a credit card number I'd gotten from my generator. And I'd hear, "Welcome to AOL. You've got mail!"

These accounts would only be good for a week or two, because once AOL got into the monthly billing cycle they'd find that Count Dracula of 12 Freddyscomingforyou Street wasn't billable and they'd close the account. So I needed a constant supply of cards to open new accounts, and that is what the generator gave me.

I also opened accounts I could go phishing from, maybe under a name like BGates or SCase, Steve Case being the CEO of AOL. I might phish fifty or sixty cards under a name like that. But meanwhile some people who got my messages would complain to AOL that this guy Steve Case was asking for cards, so the account would be canceled and an investigation would begin. Except that they were investigating SCaseAdmin, or Count Dracula of Oregon, Michigan. And when you've got millions of subscribers, it's not worth the trouble to track this kind of thing very far.

The other thing I'd do with cards I'd gotten was make long-distance phone calls, most often to voice conferences where a bunch of hackers would dial this one number to talk to each other. By now I was part of an underground society, a known person, at least by the handle I was using—iGotz. And like all the other warez traders, I wanted to talk to the guys I was hanging out with online every night.

The fact was that the deeper I got into computers, the less I had in common with people who weren't. By the time I turned sixteen there wasn't anybody in my neighborhood who understood my interests. Or at school either. Not many people even owned computers. 9X and I still hung out, but he wasn't into the programming and technology side of it, just carding and warez, so we had less and less to talk about.

The only people I knew who understood me were the ones I was talking to in the warez trading rooms every night. And after a while they became like my best friends, even though I had never actually met any of them. But not knowing what they looked like didn't make any difference. They were like me. Almost all of them seemed to be high school students, and they were all spending a lot of their lives acquiring programs and finding out about programs. Many of them were driven to understand more about the technology, like I was. They wanted to learn, and they were learning mainly from each other.

At first there weren't that many warez guys on AOL. When I first got into it there were only three chatrooms, Warez One, Two, and Three—so about 150 people all together. Maybe there was some other fringe stuff going on, but that was the core. Over time more people began figuring out what warez was, so it grew to five rooms, then six.

Actually, warez groups had been around since the creation of computers. As soon as there were computers, there was software, and everyone wanted a copy of something. That was how Bill Gates started off, by acquiring elements of the Mac operating system, changing it around, then selling it as MS Windows. The Mac OS itself had been more or less borrowed from Xerox's famous PARC developers, but that's the way the early computer world worked. You were *supposed* to trade software. There was *supposed* to be this free interchange of ideas. So when Gates did what he did, not utilizing somebody else's work, but making people pay for his product, he went against the ethic, which is one of various reasons many computer people hate him to this day.

You could say that in their own way the warez traders were continuing that early spirit. There were even people who called themselves "couriers" who did nothing but release warez. These guys would acquire programs somehow—buy them, steal them, maybe someone got some from his job, maybe they'd use stolen credit cards—they just acquired them. Then they made digital copies and released them. That was their whole purpose in life, releasing warez and subverting the capitalist system. The more they released, the more famous they became, the more other people were moved to surreptitiously give them software that they could then copy and release. I didn't understand the motivation myself; it seemed you'd be wasting too much time. Time you could be learning instead of doing something meaningless.

With so few people involved in warez trading, after a while you got to know everybody. It's actually much easier to know that many people online than in real life, because you don't have to match names to faces. All you have to do is be familiar with names, and not even regular names, handles. You're just saying hi to a word, actually.

So we'd all be interacting more or less every night and invariably someone would say at some point, "Hey, call up the conference so we can talk." One popular "conf"—conference—was called defcon, which had the same name, though with a different spelling, as the yearly hackers convention that later became popular. Our defcon was run out of somebody's bedroom somewhere in the Midwest. It became a gathering spot for hackers to chat, actually talk to each other on the phone. It gave another dimension to the contacts, and it might have seemed more secure, even though rumors were constantly circulating that the FBI had those lines tapped and monitored every conversation.

I was getting a little paranoid by this time. I was doing things on the fringe of legality, and of course crossing the line, too. So when I called defcon I never made the call from home. Or if I did, I'd use a "gateway," an 800 number that wouldn't pass along my caller ID. I'd dial up the 800 number somebody had set up, and from that number I'd call the number I really wanted. That was my method of protecting myself.

Phone calls were interesting. The thing about online relationships is that identity isn't important. You don't think about whether the person is black, white, male, female, or whatever. You know the person's handle, and that's it. At the same time you have a natural instinct to imagine a face to associate with this handle, with these words on the screen. So in the back of your mind you do have a rough idea, an image of some kind. Then you speak to this person on the phone, and you realize it's a twelve-year-old kid. And you're fifteen. And yet you've had this connection with him the whole time and maybe he knows a lot more about something than you do. You've been learning from him every night. Then you get on the phone, and his voice hasn't changed. Or you get on the phone and this guy you've been talking to online for all this time sounds like he might be Hispanic, which isn't how you imagined him at all. If you plugged into defcon often enough, all your preconceptions got shaken loose. People valued other people according to how intelligent they were, not for anything else. Much more so than on radio talk shows, for example, where callers are usually very emotional, and you're always judging them by their attitudes. Online voices are disembodied. Emotions and personalities don't exist except as they come through in the words themselves. So you concentrate on what people are saying and whether it makes sense.

But even if you had no idea who you were really talking to, a lot of socializing went on in the trading rooms. Everyone there had a common interest in computer technology. People were there to trade programs, but they were also talking and discussing how to do things. Somebody would say, "I want to make a new web page. Does anybody have Dream Graphics [or some such name]? I've heard that's really good for three-D graphics." And somebody would answer, "I don't really like Dream Graphics. I think Graphics FX is much better, because you can use add-ons." "Oh, really? I don't know about that Graphics FX." "Oh, no? You should check out this URL" (some web page); "they've got information on it." The first guy checks out the URL and comes back, "Wow, this is cool. Do you have it?" And they're back into warez trading.

I'd want to know about the latest version of WordPerfect and I'd get a detailed review from people who had been using it, with comparisons to the latest Microsoft Word. And maybe the verdict would be, "It's so much better than Word, you've gotta get it." So I would. Then I'd tell Osie, "Hey, I've got the latest WordPerfect, want me to install it?" "Yeah, sure. That'll be cool." *Ah,* I'm thinking, *I did something cool.* And Osie, of course, had no idea. He assumed I got it from someone at school or something.

Osie was working days and taking graduate courses at night, which meant I had a lot of time alone with his computer. And he was really patient with me. Even when he was home, most often he'd go into another room to study and let me stay on his computer. So I'd get an extra hour or two, even though he probably guessed I'd been at the keyboard since the moment I got home from school. Eventually he'd kick me out so he could go to sleep. I appreciated that he was letting me be there so much, but it still killed to tear myself away when he finally told me to get out.

I was glued to the screen because there was an endless stream of things to learn about. People were always talking about something I wanted to know more about, or maybe something I'd never even heard of before. That's how I first got introduced to UNIX systems and how to hack them. One day someone said, "Hey, did you hear there's this new exploit [hacking program] you can use to break into web servers?"

"What? You can break into web servers? Like change the web page?"

"Yeah, break in and change the web page."

"How do you do that?"

And for the next week everyone would be discussing hacking UNIX. And not just me and one other guy, but everybody there who was interested. And I started doing my own research, which I always did when something new got me interested. I was in school, my real school. Talking and trying things out, and listening to the conversations.

"Hey, I tried to break in and it didn't work. What's wrong?"

"What are you, crazy? You don't do it that way. What are you doing? You sound stupid. This is the way you do it."

"Oh, wow, okay. I'll try it like that."

And everybody listening in was picking up pointers. We were all enrolled in Hacking 101.

But it wasn't what you'd call a stable class. People kept disappearing. In fact, it seemed like very few stayed around all that long. Sometimes someone's parents would figure out that they were doing something bad and take away their computer access. More often, rumors would spread that people who disappeared got busted by either the cops or the FBI. The guys in the warez trading rooms had their own subculture, but they weren't elite hackers. In hacking terms they were mostly "newbies," meaning beginning hackers, unsophisticated and likely to do foolish things. Either they were breaking into websites carelessly or using stolen credit cards to actually buy things, or phishing from their real AOL accounts instead of from fake accounts. Newbie mistakes. They were caught because they were doing something stupid. So they tended not to last.

I was trying not to do stupid things myself, but when you begin getting into the world of hacking it's hard to resist this change that comes over your personality. In actuality, you've just started to scratch the surface of the technology, but in your head you have the idea that you already know a vast amount, and that knowledge has given you power you can hardly believe, especially if you're a technical person. And when you're fourteen, fifteen, sixteen years old, it's hard to keep perspective. Personally, I was getting to feel very elite, very superior. I was sure I knew so much more than everyone else around, students, teachers, principals, maybe even computer teachers. I mean, Osie, who was in graduate school and had a degree in mathematics, was asking *me* about computers. My head was swelling up like a balloon. It's really easy to start looking down on people because of the huge amount of knowledge—and power—you have. After all, there I was, age fifteen, and if I wanted to, if somebody pissed me off, I could have his phone turned off. My teacher's phone, for instance. Which isn't hard to do, but still.

The result was that I wasn't immune from stupidity, either. By this time I had plenty of real cards I'd phished. For some I had not only names and addresses, but social security numbers. On more than a few I even had their mother's maiden name, all depending on what I'd asked for and what people had been gullible enough to give me.

I had the cards, but I wasn't actually doing much that I thought would be highly illegal. Mainly I was trading them for other things. But I just couldn't resist trying to "card" at least once. The idea that I could become the proud owner of my own laptop was just too powerful to resist. If I had my own machine, I wouldn't have to rely on Osie. I could do whatever I wanted. I could be online sixteen hours a day instead of eight. If I could be on sixteen hours a day, my amount of knowledge would just shoot up like a rocket ship. The idea grew on me until I knew there was no way I could *not* have my own computer.

I didn't actually feel too bad about buying something with someone else's credit card. After all, it wasn't like I was walking into a store and *robbing* it. I was using my intellect. Some will rob you with a gun, others rob you with a pen. When you rob people with a pen, you feel so superior to all the other robbers, and so superior to the people you're robbing. And I was going to rob people with a pen. I had the ability to steal thousands of dollars' worth of stuff, if I wanted to. I had that power. Well, maybe I wasn't doing it because I hadn't figured out a safe delivery system. But knowing that I could have, if I really put my mind to it, was a great feeling. It wasn't like going into a store and sticking it up. That would be so crude, so old-fashioned.

Actually, I was so completely ignorant of these things that I assumed everybody who had a credit card must have a ton of money. And if someone was rich, it wouldn't matter that much. Later I found out that you don't have to be a millionaire to get a credit card, that your card might have a three-hundred-dollar limit. But back then, I had no idea. I might have known every bank identification number, but I had no personal understanding whatsoever of how credit cards really worked, of income requirements, credit limits, bad credit, good credit. What I knew was that you had a credit card so that you could buy lots of stuff, and that meant you must have a pile of money.

I had the laptop I wanted all speced out, an IBM ThinkPad with the works, the latest, maxed out with as much possible RAM, the largest hard drive that would fit, an active-matrix screen. I was really excited. Not only a laptop, a *free* laptop. And what's better than free?

But I knew I had to be really careful. A lot of the warez guys were getting busted trying to buy things, and word was that the Feds were cracking down hard on mail fraud. I knew the big mistake was having something delivered to your house, so the question was how to get it delivered somewhere else so you could pick it up. Maybe real thieves had figured out a good way to do that. I was wracking my brains over it.

When I finally came up with the idea, I started preparing myself. I was going to go to a pay phone so I couldn't be traced and place the order with a catalog company, using one of my phished cards. I was sure the card belonged to a white guy—the name on the card was Italian, Joseph Santorini—so over the phone I had to be this guy. So what would a white person sound like on the phone? That shouldn't be too hard. I'd take what I saw on TV and on the streets and do an impersonation. But not just an impersonation of any white guy, not a white guy per se. I'm impersonating an educated person of prestige and wealth. I'm impersonating money, someone who has money, someone influential. Someone who has a big contract coming in and needs this computer immediately to get his work done. What I want the person on the other end to hear is that this is someone who makes money. This is someone who has a job to do. I'm impersonating importance. People don't help people who aren't important, but they want to help people who are. It had less to do with white and more to do with important. Though of course money and importance also suggested white.

This wasn't going to be all that hard. I had lots of experience with the spoken word, starting in junior high school when I used to make all those speeches. Besides, I was studying acting at Performing Arts three hours a day. And my voice was already deep, like a man's. So when it came to social engineering somebody on the phone, it was just a natural extension of something I was used to already. This was easy acting. I didn't even have to do facial expressions. Pretending I was an important white person wasn't going to be a stretch.

When I finally did it, it was even easier than I thought. "Hello, this is Joseph Santorini" (in deep tones, serious). "Yes, I'm interested in ordering an IBM ThinkPad. Yes, that's correct. Model number? Yes, of course, model number . . ." A snap. I told them to have it shipped to my apartment overnight. "Yes, sir," they said. "Overnight's no problem at all."

So that part was done. Next day I took off from school and loitered around the building next door to where the guy actually lived, a nice apartment on the Upper East Side. My plan was to wait for the UPS or Fedex delivery man to show up. As soon as the truck pulled up, I would walk toward it as if I were just coming back home. I would ask, "Excuse me, is this package for Joseph Santorini?" And the delivery person would say, "Yes, how'd you know that?" "Well," I'd say, "I'm expecting it. It's my father's package, but he's not home and he told me to pick it up for him. Can I sign for it?"

That was the plan. It probably wouldn't have worked, some young black kid without any paperwork or even the right ID trying to pick up a box. But I never got the chance to try it out because even though I waited around all day, the delivery never came.

Next morning I called the catalog company, very angry. "I specifically told you I needed it immediately and it's not here! Why is that? What's the holdup?

"Sir, there's a problem with the credit card. The billing didn't go through because it was over the limit. You might want to call the credit card company."

"Oh my goodness, over the limit?" And I'm thinking, *A limit? There's a limit? How am I supposed to know there's a limit?*

Okay, so obviously having the delivery sent to the real card owner's house wasn't going to work. Even if they had come, they probably wouldn't have given it to me. And I couldn't have them sending it to my house because of the rumors of a mail fraud crackdown; I didn't know how true the rumors were, but who wanted to find out? So I figured, let me send it to the abandoned house next door to mine. I'd wait on the stoop of my house and that way I'd be able to spot any trouble before I went to take delivery.

I ordered the laptop on another phished card, set it up for a Saturday delivery, and hung out on my stoop, looking like I was just chilling. Late in the morning a UPS truck drove down the street. My heart started pumping; I was ready to make my move. But it drove by without stopping. Then, right after the truck went by, two cars pulled up and seven or eight guys jumped out, in regular clothes but with badges hanging around their necks. *Shit,* I thought, *FBI*. They rushed into the abandoned house with their guns drawn. Like a movie. It was exciting, watching the action, until it suddenly occurred to me, *Hey, they're looking for me. And I'm sitting here just a couple of yards away.*

After a minute or two they came out, hustling along a couple of homeless drug addicts who were squatting inside the building. Then one of the FBI guys said into his walkie-talkie, "Everything's okay here. How about in back?" I couldn't believe it; there were more covering the back.

While I watched, they talked with the homeless guys, then let them go. I guess they figured these guys wouldn't be into credit cards and weren't likely to want a computer.

Jesus, I thought, *that's the last time I ever try anything like that.*

9

Digital Yakuza

obviously hadn't figured out everything about this carding business; so far I couldn't do much more than make free telephone calls. But that was something, anyway. Hackers are really into calling their online friends—800 numbers, conference calls all over the country, and free is free. Besides, for guys who were doing phishing and other card-gathering techniques, things were pretty competitive. Everyone was playing games like How many credit cards can I get in a day? There was a lot of incentive to get the most, the fastest.

If you really wanted to score credit cards, you couldn't do better than break into pornography websites. Anyone who wants to use a porn site has to do it with a credit card, so these sites have huge credit card bases. If you could steal a porn site's credit card base you'd have enough to last a year, not to mention the notoriety of having done it. People were doing other kinds of hacking, too, trying to break into websites and change home pages, put up their own markings, like cyberspace graffiti. If you did something like that, your name would really be out there, it would make you famous. That's what the real hackers did.

But to change a home page, or even to attack a porn site's credit card base, you had to be equipped, which I wasn't. I didn't even know exactly what I had to do to become equipped. The reason I didn't was that porn sites used an operating system totally different from what I had on Osie's desktop. An operating system handles all of a computer's basic functions, how the memory gets apportioned, how information flows in and out, how to most effectively accomplish all the tasks the computer has to perform. Personal computers mostly run on a Microsoft operating system called MS-DOS (for Microsoft Disk Operating System) or on one of the variants of Windows, which evolved out of MS-DOS. But the websites and most other things on the internet use a hugely powerful system called UNIX. I had heard about UNIX and I wondered what it was exactly, thinking maybe it was something I should be getting into.

Meanwhile, what I was doing with computers separated me more and more from the kids in my school, but the fact was that things weren't going too well there anyway. I was immersed in technology, but my true love was acting. I had transferred over to Performing Arts because I dreamed of becoming an actor. A lot of the kids there, probably most of them, were already working actors. But my own career was hardly started and already it was getting crushed.

I was having a hard time dealing with that. I began thinking that maybe I wasn't as talented an actor as I thought I was—even though I was sure I was on the same level as my classmates. Or maybe I just didn't have that look. My manager (everyone at Performing Arts had a manager) sent me to lots of auditions and I did get some bit parts. But I had a big problem. I had developed a case of adolescent acne. If that wasn't bad enough, Grandma had also found a way to get me braces. So there I was, auditioning for commercials and movies with bad skin and braces on my teeth. And it wasn't working too well. Not many people were looking for actors of that description. Especially given the parts I was being sent up for.

I'm sure other people with braces wouldn't have gotten that much work either, but most of the roles I was going for as a young black actor were poor and street. And how many street thugs had braces? Even the commercials I auditioned for were stereotypes. "Yo, yo, yo. I'm gonna go to McDonald's and pick me up a Big Mac, know what I'm sayin'?" Targeted commercials. You couldn't be a normal person. You had to be a stereotypical black street kid. It was a little like Robert Townsend's movie *Hollywood Shuffle*, where Townsend is teaching classically trained African-American actors how to get roles. "What is it, my good man?" "No, no, no," says Townsend: "what it *be*, bro?"

These kinds of roles were a little discouraging. Musical theater wasn't going to be in my future; a nice normal kid's role on some sitcom wasn't either. But I didn't think too much about it other than wondering, *How in the world am I going to get this if I can't look the part?* Mainly, I just wanted to work. I wanted to be in front of the cameras. I would have taken anything at all.

But the parts weren't coming, and eventually my management began to lose interest. I was being sent up less and less, and then it stopped altogether. Of course in school it was the students who got the most work who got the most attention, and I wasn't on anyone's emerging talent list. My dreams of acting were crashing hard, which was making me miserable. It would have made me completely miserable, except for this other world that was pulling me in deeper and deeper, especially now that I was discovering UNIX, which was far more powerful and complex than anything I knew about.

I originally heard about UNIX in the warez chatrooms. I'd listen to some of the more advanced guys talk about it and think to myself, *This is a lot more interesting than what I'm doing. Why couldn't I learn it? Look what you can do if you know UNIX: You can hack websites. You can grab credit cards. You can get your name up.* There was also just the plain straight-out lure of technology that was begging me to come and understand it.

You could get lots of UNIX information on the net, so I started reading up on it online. I also went to bookstores. I didn't have any money, so I would sit there and read the books in the stores. Barnes and Noble was great for that. Shelf after shelf of computer books, comfortable chairs, nobody bothering you, the smell of coffee in the air. You could cut school and just sit there all day and read. You could believe you were in heaven.

At the same time I found out from some of the warez guys who were also into hacking about something called IRC—the Internet Relay Chat. IRC is a different internet technology, more specialized than the Web and harder to use, which is probably why it attracted more technologically minded types. The secret IRC chatrooms were full of people who didn't care at all about trading programs. In these places all the talk was about the technology, and how to use it to hack systems. When I started joining these chatrooms I heard names like Coca-Cola, the *New York Times,* the Pentagon, big websites people were talking about hacking, using techniques I'd never heard of. Just listening to this kind of talk sent my addiction level higher.

IRC is basically a big chat network. Now I was on it constantly, and at the same time I was absorbing all the information I could about UNIX from the Web and from my bookstore sessions. And hanging out with this new group of people, I began to see that AOL was the absolute pits of the hacking world. Really low-level. If you're hacking AOL, like I'd been doing for months now, you're regarded as dirt. IRC was where the *real* hackers, quote unquote, hung out. The intelligent ones.

The hackers on IRC were talking about vulnerabilities of different systems, quirks or holes in their security that made them vulnerable to unauthorized users. They talked about exploits, or scripts—programs you could use to penetrate one kind of vulnerability or another. A lot of these exploits were available online. If you knew about them you could download them, then use them to hack with. People who did this were called "script kiddies." Script kiddies weren't newbies, beginners, but they weren't high-level either. Above them, on a completely different plane, were elite hackers who studied systems for their vulnerabilities, then wrote their own programs to exploit them. They were the ones who understood the technology that creates programs and the programming languages. They lived in a different world—one that I was just barely glimpsing now, that I'd never considered entering before, but which I could aspire to, even if I was going to have to start from the beginning.

The beginning was finding all the online documentation that had anything about hacking UNIX, and that's what I was now reading full time. The documentation told me that there are various types of UNIX operating systems, just as there are different variants of Windows—Windows ME, Windows NT, Windows 2000, Wndows XP, et cetera. UNIX systems are the backbone of the internet. Even going back to the beginning, before the internet, long before I was born, most of the banks, telephone companies, and such were controlled by UNIX-like operating systems. So there was this whole long history of UNIX and its variants. And for each type of UNIX operating system, the documentation said, there may be several default passwords that are built into the system for one reason or another. If you knew these passwords and their associated user names, you could get on.

So at first I would connect to servers of companies or organizations I thought might be interesting. I'd telnet in and I'd see, for instance, a banner that read UNIX System V Release 4, Login—computerese for "log in." I know (because the documentation told me) that "System V" meant I had found a SunOS machine, a computer running Sun Microsystems' version of UNIX. And the documentation also told me which user name/password combinations might give me access to that kind of machine.

I'd key in the first user name and password, the second, the third, I tried them all, hoping that one of them would let me on. Sometimes I didn't even worry about which system a machine might be running; I just tried all the default names and passwords I knew. It was an archaic and horrible way to hack, the equivalent of randomly choosing apartments to try to break open. Like if you look for an apartment door that seems interesting, and you try to break the lock open with an inadequate burglar's tool that may work one time in a hundred. Now, while you're doing this everyone on that floor is going to hear you, and if the person is home, he'll hear you for sure. I was doing the equivalent of brute-force apartment robbing. Which made it very dangerous. But at that point, this was the only method of hacking I knew. Most times, of course, the default login didn't work and I'd see on the screen Login incorrect. When none of them worked, I'd go off and try another computer.

Some UNIX systems, such as IRIX, an operating system made by Silicon Graphics, were especially easy to break into. IRIX was very popular among hackers for having two or three default accounts that could be used without any password at all.

IRIX was the greatest. If I connected up and was greeted by IRIX version 4.1 I knew just what to do. When I was prompted for my password, I'd key in lp for line printer, an application that runs on this server and for some reason has a default account with no password. I'd hit enter and see something like Welcome to the Worldwide Security and Protection Corporation. Lp is a really low-level account, though, which meant that I wouldn't be able to run many of the programs on the server. But it did let me look around to see what was there, including the password file, which I could now copy onto my own machine. At first I wouldn't be able to do anything with the passwords, since they were all encrypted. Later, when I acquired a password cracker, I'd run them through in hopes of cracking the root password.

Looking around a machine you've hacked into is the same as if you were actually sitting in front of that particular computer deciding which programs to run or which files you want to use. A server will have various directories, just as you have various directories or folders on your home PC. And each directory has something in it, whether it's programs or documents or pictures. People tend to name their folders, their directories, according to what's inside them. So you go through the directory names to see what might be of interest. A directory titled "Secret Documents," for example, might grab your attention. If the account you're using is at a high enough level to give you access (a root account will give you access to everything), you can run a command that gives you a list of all the files in that directory. And if you see anything interesting, you FTP it back to your own machine, FTP being the protocol that enables you to transfer files from one machine to another.

I couldn't do that kind of thing at first; I was just looking randomly for things that struck my curiosity, looking for the sake of looking and learning. I wasn't breaking in to destroy or steal anything, I was breaking in, first, to learn how to break in, and second, to learn how to use the different kinds of machines I was breaking into. I'd be on a machine I'd gotten access to and at the same time I'd be searching the Web for commands I could run on this machine's particular operating system. If I found something I could use, I'd come back to the machine I hacked and I try to execute the command.

That's how I created my first back door. I had connected to an IRIX machine, using a default account to gain access. Then I gained root. And while I was on I found a document that showed me how to edit the password file, for example, by adding an additional account. If I want to add a user to this machine, the documentation told me, run the command "adduser." I run it, and I see on the screen, Enter new user name you would like to add. I enter "jdavis," and I get back, Enter password for user name jdavis." I type in "jdavis," to keep it simple, and I hit enter. Now I have an account on this machine called jdavis in addition to the default account that I used to log on in the first place, two accounts I can use in the future to get back onto this machine whenever I want. If they kill the jdavis account because they realize it's not supposed to be there, I still have the lp account. If they realize somebody is using the lp account who shouldn't be, well, I have the jdavis account. So I've just learned something new. I now know how to create a back door; at least, one kind of back door.

Being on IRC was like being in a school for breaking and entering, at least being on the hacking channels I was spending my time in. Everybody there was sharing his or her knowledge of exploits—what they were, where to find them, how to use them, how to cover your trail. And people were always in the chatrooms, so you could be talking to them even while you were trying things out, like on-the-job training. Every day, I was racing home from school so I could start hacking. I couldn't wait to use the latest exploit, to see how many machines I could break into with it, to look around all these new environments I was finding myself in.

I'd be out trying these things, feeling a surge of power and elation from breaking in, from finding a root password and "owning" a machine. And at the same time I'd be going back to the rooms so I could tell people what I was doing and brag some: "Listen, listen, I just used the PHF exploit. I'm on a big system here, you ever hear of a school called MIT? Yeah, I got in through some student's computer. Yeah, I've got root. Now I'm going to . . . hold on, hold on, I think their administrator just came onto the machine, I have to clear my logs." Or, "Hey, I just broke into this machine, what should I do now?" Because I'd never seen that environment before.

I would hack a website and I'd be like a sloppy thief. Like breaking into someone's house, looking all over the place for the jewelry, grabbing it, running out of the house leaving fingerprints, open doors, everything. But the IRC guys would tell me, Look, once you get into the house, you have to know certain things, like where the jewelry is. It's always in the bedroom, so go there. And it's usually in drawers, in underwear drawers, so go there. And sometimes people leave extra keys in the house. So take them and make copies. And if you're neat, you can come back some other time. From the IRC guys and from all the hacking documentation that was accumulating on the internet, I started learning how to gain access to additional systems once I got onto the first system. How to always hack from a remote machine; that is, from a machine I had already hacked into, not from my home computer. How to clean up after myself so I couldn't be traced.

Hacking all of these machines gave me an incredible feeling of mastery, but I was also losing that superior, arrogant attitude I used to have when I first got into hacking. Now I was hanging with guys who were just as into the technology as I was, many of whom knew lots more than me. So that helped make me more humble. Gradually I fell in with a group who had more or less the same interests, lived in New York City, and played in the same circle. Most of us had met each other in the AOL days and had evolved together into the UNIX environment. I had talked to two or three of them on the phone, and one day they invited me to get together with them at a 2600 meeting, their usual monthly meeting place.

A 2600 meeting is usually where newbie hackers go to meet people who are doing the same thing, but others in the hacking world stop by too. The name 2600 comes from a magazine that's popular with hackers and phone phreakers—people who specialize in hacking the phone system. But its origin is in the famous moment in hacker history when John Draper discovered that the prize whistle in boxes of Cap'n Crunch cereal emitted a tone of almost exactly 2600 megahertz, which was the tone AT&T used to terminate long-distance calls. Draper found that if you made a long-distance call, then blew the whistle into the mouthpiece, the AT&T billing system would think the call was terminated, even though you were actually still connected. Which meant your call was free from that point on. This led to a whole cult of people who devoted themselves to playing games with the phone system, like calling a phone in the next room by routing the call around the world—all for free, of course. Anyway, ever since, Draper's been "Cap'n Crunch" to the hacker world and "2600" has been a kind of hackers' symbol.

For anyone just getting into hacking or phreaking, 2600 magazine is a great way to learn. It contains all of these informative articles about things like telephone-hacking codes and other hacking techniques. And at the back of each issue it publishes information about where and when 2600 meetings are held in each major city.

The one in New York was at the Citicorp Building. At one time it was located inside the big atrium lobby, but then Citicorp banned them from the premises. Maybe the hackers were doing things they shouldn't, like some guys would play with the security guards' radios by broadcasting on their frequencies, or maybe the FBI told Citicorp they shouldn't have hackers in the building. Anyway, they were banned, so the meetings moved outside, or between the two sets of entrance doors to the lobby, depending on the hackers' luck with the guards.

I had gone to a meeting once with 9X, who had introduced me to 2600 magazine. "Hey dude, you got to check this out. It's got phreaking stuff in there and everything. It's really cool." So one day we decided to go to a meeting. It was a sight. There were fifty or sixty people there, running around with walkie-talkies, phones, beepers, laptops, documentation, credit card readers (that read the embedded information on magnetic strips), all types of hacking material. In one place a bunch of AOL kids were trading stuff. In another someone was showing a new program he had on his laptop to a little circle of guys. Someone else was handing out papers showing how to write credit card readers, showing them how he had made his and answering questions.

A lot of them looked like nerdy young kids, but there were also old guys from the old ham radio days who were in their fifties, who were now into other things, like hacking computers. Some of the old guys were wearing plaid shirts, but most of the young ones were in black, like goths, some of them with their faces pierced. As if they were trying to look badly antisocial, because they believed that that's what a hacker was supposed to look like. Not that I knew; at that point I had never seen another hacker except 9X, who looked like a skateboarder.

There were clusters and conversations with five or six or eight in each little group. There was lots of activity, lots of bouncing from group to group, lots of electronics. I thought it was amazing, exciting. But I had to maintain my elite demeanor, which meant that I couldn't talk to anyone. I wanted to, but I didn't know anyone, so it was easy not to make the effort. And 9X was like, "Aw, man, these guys are so lame." He's smoking a cigarette, trying to look cool too with his orange hair. He and I are hanging on the side and 9X is blowing smoke into my face, talking about how lame these kids are. And I'm like, "Yeah, what lamers." And meanwhile I'm thinking, *Hey, I wonder what those guys over there are doing?*

In a way I did think it was lame, because it was a bunch of twelve-year-old kids running around. And at first I thought the old guys must be Feds, though later I identified them as old Lefties, the ones wearing plaid shirts. They were the same guys you see hanging around hackers' conventions today, telling people, "You have to join the revolution. We need hackers like you. We can bring you up to the camp, train you to fight." Old hippies who are still antisystem, antigovernment. Guys who naturally gravitate toward this type of environment.

The impressive thing about that meeting was the frenetic energy, lots of kids with this excitement and curiosity about the technology—which was what I felt too—but also an edginess, like they wanted to appear bad but were too wrapped up in what they were doing to actually *be* bad, at least for the moment. The area of this big indoor plaza where they were running around was just crackling. But most of them were really young, and I did feel self-conscious and out of place since I didn't know anyone. So after that time with 9X I didn't go back—until my new friends invited me down.

When I got to the Citicorp lobby it was like time had stood still. This was maybe seven or eight months after my first visit, but there were exactly the same people doing exactly the same thing. I stood around for a while watching until I noticed a couple of guys glancing at me, trying not to appear too obvious. One or two of them I recognized—they had sent me digital pictures—especially one with red hair and freckles, so it was kind of hard to miss him. But they had never seen a picture of me, so they're catching my eye and they're like, "iGotz, is that you?" "Yeah, Zeed?" "Yeah. This is so and so, this is so and so, this is so and so. This is the crew." So now I was part of a crew.

We didn't stay at that meeting. We never stayed at any of them. We'd go there, hook up, talk a little, then we'd leave and go eat at McDonald's or go somewhere to hang, maybe an internet café where we'd show each other new things we had learned.

Our primary goal was not to hack the world, at least not together. First and foremost we were a hangout crew. We were all hacking away on our own, and sometimes we would tell each other about it. But when you start getting good, you begin trusting fewer and fewer people. You become wary, even with your friends. Because you understand how easy it is to get caught and you know that some of the things you are doing might be felonious, *are* felonious. So you tend to say less, and very little directly. The way it would come up would be, "Oh, did you hear about this website that got hacked?" And one of the guys might say, "Hah, they've been owned. Trust me, they've been owned a long time before that happened." And everybody would know he was saying that he had hacked that site way before the recent hack. And he might still be in that site.

We all felt this strange contradiction. We knew we needed to be really tight-mouthed about what we were doing. But we also wanted to be famous, to get that recognition and respect. At the time most of us thought that the good hackers were the famous hackers (I eventually learned better). Everyone admired the hacker who hacked a major website. What you actually did to a web page once you got root would be up to you, depending on how artistic or subtle or clownish you might be. Maybe you'd create an elaborate web page that duplicated the original but had minor logo or name details changed, like Amazone.com or Barnes & Ogle. Or maybe you'd change the entire web page and paste a naked lady up there, maybe Madonna, with the words "Hacked by Moby" under it, or

"Moby loves Madonna." Next day the tabloid headlines would read HACKER MOBY SENDS LOVE LETTER TO MADONNA. And now Moby would be famous. There was even one website, www. attrition.org, that archived all the hacked websites. You could go there (or to its successor, www.alldas.de) and check them out.

But whatever the hacker actually did to the web page, if it was a big site all of a sudden he'd be "one of the best hackers in the world." When you hack a major website you go from zero to hero in a matter of hours. One day no one knows your handle; the next, you're in every newspaper across the country.

By now it was months after I was introduced to IRC and I could hardly remember why warez trading had ever seemed so exciting. I was still mainly a script kiddie, downloading and using other people's exploits, but already I was getting into coding, figuring out how to write my own programs, in my "let me figure out the underlying technology" phase. The word everybody used for hackers who knew what they were doing was "elite." I was definitely getting more elite, more expert at using hacking tools, and more knowledgeable about how to protect myself.

Hanging with my new crew, especially two of the guys, who worked for internet service providers, helped. These guys had access to extremely powerful machines and had done in-depth UNIX hacking. But everyone had ability and there was constant cross talk about what to do and how to do it. The social part also had an interesting side, because these guys were like any other group of hackers. They were people drawn together by the technology who might have been very different from each other in other ways—like calling a hacker on the phone and finding out he's a twelve-year-old kid with whom the only thing you have in common is an interest in web design.

That led to some strange happenings, at least for me. The red-haired, freckled kid I had noticed first at the 2600 meeting was about as far from my experience as I could imagine. His family wasn't just well-off, they were rich. He lived in a huge apartment on Madison Avenue filled with antiques. I had never seen anything even approaching that. He went to a high-class prep school and would come to the meetings right from school in his blazer and tie, or else he'd be wearing shorts, tennis shoes, and a polo shirt. Everything about him smelled of prep, or what I considered prep. He talked about flying to Paris for spring vacation just to hang out with his relatives who lived there.

Once I was at his apartment helping him with something he was working on when he suddenly said, "Hey wait, I want you to meet someone," as if he wanted to introduce me to a member of his family but had just forgotten until now. Then he went out of the room and came back a minute later with his maid, who was black. And he said, "This is my maid." This woman and I were standing there looking at each other. He was standing between us waiting for some kind of black people connection to happen, now that he'd brought us together. This was not a bad kid; I'd never picked up any hint of racism on him. But he had an odd perspective on black people. He said, "This is my maid." I'm standing there look-ing at this woman, she's looking at me, and we're both thinking, *What in the world?* And I'm like, "Uh, yeah, nice to meet you," thinking, *So? Okay, she's your maid. Now what?* I guess he was proud, like, "So, see, I know other black people too, not just you." He seemed to need to demonstrate something. He got back from France and he said, "Yeah, I just came from Paris and I met this black girl on the plane. And she was *so* hot. We were kissing and everything. On the plane. Me and this black girl. A *French* black girl." And I'm like, "Hmmm, cool." I guess he liked women of color.

One day after a 2600 meeting we were sitting around, figuring that we needed a name to call ourselves. We knew we were good, maybe the best group in New York City—which maybe wasn't saying that much, because there weren't many hacking groups in New York City at that point. We were going to be the next Masters of Deception, the legendary crew of hackers from six or seven years back who had burrowed so deep into the AT&T system that when half the country's long-distance phone system crashed they didn't even know themselves if they had done it. Except of course we weren't going to jail like they did. Anyway, we needed a name. The one we finally decided to go with was Digital Yakuza.

I was hanging with Digital Yakuza at the end of my second year in high school when my horizons shifted 180 degrees, by accident. Totally unexpectedly my path got changed around, although it took a while before I was aware that it had. Out of the blue, a system administrator offered me a job.

What happened was that I was hacking one Friday, still using Osie's computer. Fridays were good, because Osie was hardly home then, so I was hacking away, playing games with myself, thinking, *Okay, today I'm only going to break into ISPs.* Actually, I was also looking to change internet service providers, to get away from AOL, so here was a chance to kill two birds with one stone. I could choose likely ISPs, then test their security by trying to hack them.

The first thing I did was go to a website that lists all internet service providers for people looking for new service. Of course, it turned out there were quite a few located in New York, so I thought, *Okay, why don't I limit myself to the city?*

One company I checked out was vulnerable to the PHF exploit, a hacking program that had been around for some time. On the spur of the moment I thought, *You know, why don't I just call the administrator and tell him he's got this vulnerability in his system?* Ordinarily, I'd never get involved with anybody I was hacking. But this guy was in New York, so I thought, *Why not?* If worst came to worst, I couldn't go to jail for telling him there was something wrong with his system. Besides, I felt this sneaky desire to be admired for what I could do, and who better for that than the system administrator of the site I was hacking?

I called up the company's office and asked to speak to the person in charge and I got this guy on the phone.

"How can I help you?"

"Yes, I was just looking around and I ran into your website, and I noticed that you are vulnerable to a particular attack method. I thought you might want to know about it."

"Really? Sure. Tell me about it."

"Well, you're vulnerable to the PHF exploit."

"PHF? What's that?"

"It's a program that comes with your web server and allows subscribers to view files, but it's really being exploited now to allow outsiders to view your password file. If you'd like, I can send you something to show you what I'm talking about."

"Yeah, that would be great. But tell me, how come you found this?"

"I was interested in using your service and I decided to test it out a little bit and I noticed."

"Oh, I see. Listen, I really appreciate this phone call. I'm going to look into this and try to get it fixed. By the way, what is it that you do? What kind of work?"

"Well," I say, in my deepest voice, "right now I'm a student, but I do computer security and web page design."

"Oh, really? Would you like to come down here for an interview sometime? We're looking to hire people."

"An interview? Sure, I could come down for an interview," already thinking, *What kind of clothes do I have that I could wear to an interview?*

10

Hacker

All I had were my school clothes, blue jeans and T-shirts mostly, so Osie loaned me a dress shirt, tie, and slacks and off I went into Manhattan to interview at the guy's office, which was right off of Twenty-third Street. I can imagine what he thought when he saw me, because I don't think he was expecting who I turned out to be. I think he was expecting a college student majoring in computer science. But if he was surprised, he hid it pretty well.

First thing we did was walk around the office, and I was the one who was surprised. I had never been in an office like this before. I had hacked into plenty of ISPs but I had never actually been to the offices of one, so I didn't know what to expect. This company had a pretty big subscriber base, a couple of thousand users, but it was run in a one-room loft, with the server room, where all the important computers were, in a kind of big walk-in closet. Inside that room it was warm and dark, glowing with blinking lights from

modems and multiple servers sitting on racks. I'd never seen anything like it; all those blinking lights made it look like something
out of a science fiction movie. "Wow," I said, I couldn't help
myself. "Wow, this is cool," which probably gave away any hope
I had of impressing this guy with my sophistication.

Sasha (I'll call him that) introduced me to his two employees,
one of whom was his girlfriend. All three of them were Russian.
Then he asked, "What can you do?"

"I can write HTML. I know FreeBSD. I know what the different UNIX operating systems are. I know basic system administration. And I'm really good at security," by which I meant that since
I was good at breaking security maybe I could also be good at
improving security. "I can do graphic design too."

"Great," says Sasha, "we can use another web page designer.
And we can use someone to help with security. We don't know
too much about that."

He explained what the company did. They were an internet service provider, but they also did web page design, as well as HTML
and CGI programming, plus UNIX administration. It sounded
like a great place to learn a little bit about everything. Then he
hired me on the spot. I was the fourth employee. Summer was
coming up, so I could start off full-time, then go to part-time when
the school year began in the fall. The pay was $8.50 an hour,
which would come out to about $5.50 after all the taxes, so would
I mind, he asked, if he kept the three dollars and paid me cash? I
hardly heard him. They were going to pay me for working in a
place like this? When could I start?

When I went back home and told Grandma I got a job, she did
an immediate 180. Before that, what I heard every day and every
night was, "You're spending too much time on that computer!

You need to go outside and do something! You need to get some fresh air! You need to get a job! What are you doing on that computer all the time, you playing games on that thing? How can you play games all the time?"

That lack of family support had really gotten me down. Sometimes I felt that no one understood me, that the whole world—except for my online friends—was against my desire to understand this technology. But now all of a sudden it was, "Ejovi got a computer job! Ejovi sure knows what he's doing with those computers. Ejovi really knows all about that computer stuff."

The thing was that Grandma and the rest didn't have any idea what I was up to. Once they saw I was actually doing something productive and that I could make money off of it, I started getting all the support in the world. Osie had actually been supportive all the way along, no matter what he might have said from time to time. He was the only person who never really gave me a hard time. He was curious about what I was doing. He asked, but I don't think he understood exactly. I told him I could trade programs with my friends. He didn't know that the act of trading programs was illegal. He just thought, *Okay, he's trading programs with his friends, good.* I was also trying to start a design business, so I was creating web pages. So he thought that was what I was doing, graphic design, web pages, building my skills—which *is* what I was doing. But the other part he didn't get.

As for me, I felt like I had finally accomplished something, that all the time I spent on the computer was paying off. I was happy. This was my first official money-paying job, even if I did have to give the guy a kickback. And I also got a free account, which meant I could browse the Web all the time, from my own computer.

Office hours were nine to five, but with so much to look at I'd usually stay late. But one night around ten o'clock, not too long after I started, the door opened and in came the boss with my co-worker, his girlfriend. "Hey," he said, slurring his words a little, "what are you doing here?" They looked somewhat drunk. "Oh, I'm getting some work done, doing some research online. But I'm about to go." I was obviously interrupting something.

Another time I came in early and found him sleeping on the futon. That's when I realized that my boss was living in the office, and that his girlfriend probably slept over most of the time. It had never occurred to me that a regular computer company might be run out of a loft with the owner sleeping on the futon. But there he was, sleeping, and then taking showers at the University of Pennsylvania Alumni Club, as he told me later. I guess I had imagined something more like a regular corporate environment.

This job seemed like something where I'd learn a lot, doing a lot of things, which is exactly how it turned out. I was doing system administration, programming, a little web design, and, of course, security. And one day I was checking the servers and I found that one of our subscribers had hacking tools in his directory—exploits he could use to break into systems. So I took the opportunity to start a discussion with him and get to know him a bit.

His name was Grim and he turned out to be a talented guy. He had been hacking longer than I had and he knew about things I didn't, like BugTraq. BugTraq is probably the longest-running security mailing list on the internet. Back then it was a "full disclosure" mailing list, a great place to find new exploits. Hackers would post new vulnerabilities that they had found, and exploits they had written, new ways to break into systems. It was supposed to be a security mailing list, but it really functioned as a hackers' mailing list for exploit code. At the time there were probably about five thousand subscribers. Now there are sixty or seventy thousand. BugTraq has grown so much that now it's a commercial company, an important mailing list for security professionals. It was a security list back then too, but most people used it as a place to gather exploits to help them in their hacking careers. Which was the reason I subscribed right after Grim introduced me to it.

Grim knew not only BugTraq, but also similar lists, like 8GLM. He had exploits I didn't have, and was adept at using them. One day I was running a security check and I discovered that Grim had hacked our server. He had gotten new exploit code off the Bug-Traq list, compiled it, and tried it out on us. The particular exploit he used was what's called a "buffer overflow," a way of sending illegitimate instructions to a legitimate program that's running as root, and thus telling it to do something it's not supposed to be doing. A hacker using a buffer overflow can execute root commands through the program he's managed to subvert.

So there Grim was when I found him, sitting on our server, with root privileges. I was so impressed, I just had to show my boss. "Look at this," I said. "Look what he did. Isn't this incredibly cool?" And my boss said, "No, this is *not* cool. This is what I'm paying you to prevent!"

Sasha was a graduate of the University of Pennsylvania. He had studied computer science there and sometimes if we were just sitting and talking he'd tell me about what he had done there, about his professors, his courses, the kind of computers he used, and generally what a great school it was and how much he had learned. Before that I may have heard of the University of Pennsylvania, but I'm not sure. Places like that—Pennsylvania, Harvard, Yale, Princeton—I only knew about vaguely. I knew about them to the extent that I understood I'd never be able to get into any of them. In terms of money and grades they were simply beyond anything I might have thought about. But I was curious, and all of my boss's talk got me to thinking that even if I couldn't go to the University of Pennsylvania, I could at least look around its systems. Why not hack in and see for myself what type of stuff the people there were doing?

The more I thought about it, the more interesting it seemed. I didn't have to write away for their brochures and course lists or go take a tour; I could just break in. Besides, I wasn't actually that interested in what the place looked like or what the courses were. What I wanted to see was their computers, and who could give me a better tour of those than myself? I knew that universities have the most advanced machines because of the research they do, especially the really great universities. They had computers I'd never be able to see or touch in my lifetime. On my own machine I could crack about twenty passwords an hour; on theirs I could probably crack a thousand.

I was sitting there at my desk fantasizing about these things. I didn't know about mainframes or supercomputers, I didn't know what kinds of systems they had or what they ran on them. All I knew was that they were much bigger, faster, and better than anything I would ever get near. I was in a reverie about it, trying to imagine what such incredible systems would be like. Daydreaming. Until one day when I was kind of bored I thought, *Okay, I'm not doing anything else at the moment; why don't I just hack them right now?* So I did.

The first thing I did was run a domain name service lookup, which allowed me to view the computer addresses at the University of Pennsylvania's dormitories. There I found a student's machine that was running a web server I could get onto using one of my favorite exploits. So I ran that, got on and got the password file, which was encrypted, of course, but I put it through my password cracker. When I cracked the password and logged onto the student's machine I discovered he was running a Linux operating system, which told me he was probably pretty sophisticated, and when I looked around his files I found out he was a computer science student.

Once I gained root access to the student's machine—this probably took me about ten minutes—I downloaded sniffer source code from the internet and compiled it; that is, I translated the code into language the machine could use. A sniffer is like a telephone tap; it allows you to monitor all the traffic on a network. You can watch everything, and since the student's computer was on the university network I was watching all of the information going back and forth over their wire, although I was only really interested in capturing passwords and destinations.

When I had grabbed a bunch of passwords for the university servers, I logged onto them and hacked them. While I was doing this my boss was sitting across the room at his desk, but I wasn't too nervous about getting busted. I figured that if he came over for something I could just kill my screen. Or maybe I'd even show him what I was doing. He should have been pretty interested, since Pennsylvania was his school. Or maybe not. Anyway, he looked busy and I wasn't worried. I was too busy hacking into the servers from the student's machine, then setting up my sniffers on the servers I hacked. If I keep going like this, I thought, I can probably own their entire system. Then I could take my time and really explore what they had down there. This was very exciting, a whole university with a top computer science program. They had to have things I hadn't even dreamed of, and I was going to see them all. By this time I wasn't even aware of my boss anymore; I was lost in what was going on in front of me.

What led me into hacking the university was the seductiveness of the technology, wanting to know about it. But actually there was more to it than that. A hacker is like what I imagine other voyeurs are. You find yourself a part of things that aren't yours, in relationships or events that don't have anything to do with you, which you are a stranger to, but suddenly you're connected to them. In a way it's about being someone other than who you really are, who *I* really was.

There isn't very much difference between breaking into a university's machine and an individual's. If I break into your machine, all of a sudden I'm you. I'm seeing the same things you're seeing. I'm viewing the same web pages you like to watch. I'm reading your personal letters to your daughter. And I feel this level of intimacy with you, because I'm in this special place I would never have access to otherwise. I mean, you would never tell me your most intimate feelings; but if you keep your diary on your computer, then I'm there with you.

So if I'm breaking into Pennsylvania, I'm feeling a certain intimacy of being a part of the university. That was the real lure. Why else would I want to break into a school? Who wants to? What thief says he wants to break into a school so he can read its books? But it's kind of like how some people break into the underground tunnels of schools, to go places other people haven't gone, places other people don't know about. To give them a sense of knowing, of intimacy with the environment. They don't want to put a bomb down there, they just want to *be* there. The closest I could come to being a part of those schools was to break into them. And then I felt like I was really there—I wasn't just a kid from Bed Stuy, in a way I was part of the university.

Over the next few days, whenever I had any free time I spent it looking around the ten or so machines I had managed to break into. Then one day I was in the system and suddenly my server connections shut down. Then my point of entry, the student's machine I had used to hack the servers, shut down. I saw on my screen connection closed and I thought, *Hmmm, that's probably not good.* I was racing through the steps I had taken, to see where I might have made a mistake—I could see some of them almost instantly, when the phone rang and my boss picked it up. I was thinking, *How could I have been so sloppy? And how can I cover myself?* I was running through the possible scenarios when I heard him say, "Hold on just one second." I could tell from the sound of his voice that something was wrong.

I was looking across the room to where he was sitting, and he was saying into the phone, "Oh, really? That's interesting." Now he was looking over at me. I was glancing at him, but trying not to. He said, "Listen, hold on, will you?" He put the phone down and said, "Ejovi, um, were you doing anything with the University of Pennsylvania?"

"Uuuh, kind of. Why?"

"Because I'm on the phone with their security administrator right now, and he's telling me that someone broke into his machine. Would you know anything about that?"

"Well, I used this one exploit. I was just looking at it; I kind of wanted to see if they were vulnerable."

"Wait a minute, hold on. Let me put this guy on the speakerphone."

He put the security guy on the speakerphone and I heard, "Look, I'm not fooling around with you. I'm gonna call the FBI right now if you don't tell me what you did, which machines you broke into, and how you did it."

Most of this was probably a bluff. Whenever I thought about being caught and what might happen I figured that first of all I wasn't stealing anything or causing any destruction when I broke in. Second, I was underage. I had just turned sixteen. The worst they could do was slap my wrists, give me community service or something, right? But now that I was hearing FBI, I got scared.

"Well, uh, I broke into this machine and, you know, someone else told me about it, and they broke into those other machines, so I don't know about that. But I can tell you how they broke into them."

"All right. Tell me everything you know right now."

So I spent most of the next two or three days on the phone with the University of Pennsylvania security adminstrator, explaining everything I did to their systems. My boss was pretty upset. I was sure he was going to fire me, but he didn't, though I wasn't clear why. Maybe because I admitted it, maybe because he realized I wasn't malicious, or maybe because he figured I was young and didn't understand the consequences of what I was doing. After all, I had broken into *his* system and had gotten a job out of it.

This Pennsylvania guy wasn't offering me any jobs, though. He really grilled me. I had to go step by step through the process, which actually wasn't all bad, because I was able to understand exactly where I had been sloppy, and why they had caught me.

The moment I started my explanations, in fact the moment I had heard my boss on the phone with this guy, I realized my first mistake. I hadn't cleared my logs, which is a fundamental hacking rule. Maybe I was too excited or something, but not doing that was just plain sloppy. I knew that whenever anyone accesses a system it's logged into a database. Every time you dial up AOL to check your e-mail the AOL database shows that you connected to the network at this date, this time, and that you signed off at this date, that time. Whenever you go onto the internet, it's logged. Every time you connect to a UNIX host, it's logged. All your comings and goings, with dates, times, and addresses of who's sending and who's receiving; they're all logged.

Once I got root access to the student's machine, I had the ability to delete the log-in data. I could have modified the system so it wouldn't show that I had been on it, I could have removed their ability to see me. Nowdays it's possible for administrators to check a system to see if it's been modified, but in 1996 not many people were using tools to watch for system changes. Back then the technology for it was just coming into use.

But I hadn't cleared my logs. Not only that, I had hacked the student's machine from my desktop at work. I hadn't used a cutout (an intermediary machine), or several cutouts, like I might have. So the IP address that was logged in was my company's address, my desktop.

"Okay," the Pennsylvania security guy said after I told him how I had gotten onto the student's machine, "what did you do then?" So I told him how I had run the buffer overflow to get root, then downloaded the sniffer, compiled it on the student's machine, and started capturing passwords. What I didn't have to tell him was that instead of deleting the exploit code from the student's machine after I had used it, I had left it there, where it could be found by anybody whose suspicions were aroused and started looking. I had also left the sniffer code on the machines, another telltale sign that someone was up to something.

Finally, there was one big mistake that I hadn't suspected. "You'll never believe what happened to me yesterday," I told the Digital Yakuza guys while we were having burgers at McDonald's. "My boss was telling me about the university he went to? How it was so good and the technical people were really smart? So I decided to hack it, and next thing I know some system administrator caught me and I'm getting a phone call from him." They're all ears. "So I was sniffing passwords and I got about six hundred accounts, but I only told him I had gotten thirty or forty. Anyway, everything's okay now. I still have access if I want to go back, but I don't want to take the chance of getting fired or arrested or anything." Very nonchalant, like no big deal. Like, yeah, they caught me, but I still outsmarted them.

"Well," said one of the guys, "you do know that when you run a sniffer you're setting the machine to promiscuous mode and it puts a message on the monitor."

"Well, yeah, of course I knew that. I just didn't think he'd be looking at it." Which probably sounded a bit lame. No doubt. Because although I knew that when you run a sniffer it sets the computer on promiscuous mode, I did not know that a message comes up on the system broadcasting `set in promiscuous mode`—"promiscuous" meaning that the machine is looking at all of the information coming across the wire, instead of just the information being addressed to that machine specifically.

These days there are better sniffers that don't run alert messages. Then there weren't. You just had to hope the administrator didn't look on the monitor. Of course, many machines are just sitting in the network doing whatever they're supposed to do, without anybody having any reason to look at the monitor; the monitor often isn't even hooked up. But if you're breaking in, you take your chances. At least back then you did.

Now that I know this I can see the whole thing from the security guy's perspective, how he caught me. There I am, deep in the university system for days, and nobody's noticed a thing. Until one morning an administrator is checking his monitors and he sees a message on the monitor of one of my hacked machines: PRO-MISCUOUS MODE. He sees that and his heart starts racing a bit; it's a red flag. He starts looking into it. Maybe he's on someone's trail here, because why would a machine be set that way unless someone's running a sniffer on it, and who would be running a sniffer except someone who ... So he logs into the server and does a "w" command—a "who?" to see who else is logged in. Hmmm, there's someone he doesn't recognize, someone not authorized, and now he knows he's on the trail, because that's the person who set the sniffer up. And this person, whoever he is, has conveniently left his logs on the system, with his internet (or IP) address, which the administrator goes to and finds is that of a student living in the dorm, a computer science student, which makes sense, with a Linux machine. Okay, so, what now? Well, let's look around in this student's directories to see whether there might be some clues there. The administrator feels like a cop searching an apartment for evidence. And he opens a directory and finds not just a clue, but a smoking gun, along with bloody knives and gloves. Sitting there in the directory he sees a sniffer program, yes, and a buffer overflow exploit, which the perpetrator has inexplicably not gotten rid of: exploit.c and sniffer.c. And something else. He checks the logs and finds a strange account logged in from some unknown outside IP. So, it's not the student, is it? It's some-one who hacked the student's machine and is using it as a plat-form to hack the servers. The student doesn't even know what's going on, it's this other guy, who oh so conveniently left his address: desktop21.somecompany.com. An internet company, no doubt with a website, which he goes to—after shutting down the student's computer, of course—and finds (along with anything else he wants to know about somecompany.com) the company's telephone number. Which he dials, and begins talking to the administrator, who says, "Hold on just one second." And then a muffled, "Ejovi, um, were you doing anything with the University of Pennsylvania?" like the guy got nervous or something and for-got to put him on hold but just has his hand across the telephone mouthpiece. Which sounds to the administrator like maybe he's got the guilty party.

11

Facts of Life

In September, Performing Arts started up again, and before long I could see that my work at the ISP was getting in the way of what I should have been doing in school. My acting career still wasn't going anywhere, but I didn't want to let school slide, so I decided to quit and try to buckle down for a change. Also, by then the five-fifty an hour wasn't looking nearly so good as it had before.

Maybe it was really that the ISP wasn't interfering with school, it was interfering with hacking. Between school and work I didn't have that much time, and Osie's computer was just sitting there at home, many times all night long, almost as if it were calling me.

By now I was getting smarter about breaking into machines and covering my tracks. And for some reason I just seemed to have this need to break into as many as possible, almost like something was driving me to do it. I was obsessed.

When Osie wasn't home I'd be at the computer all night. I'd hack into a machine, set up a sniffer, and hack any machines I saw on that network. If the sniffer gave me twenty passwords from twenty other machines I'd hack them all, and each might give me twenty more passwords to twenty additional machines, which I'd then hack and set up sniffers on.

Hacking internet service providers was the best. A big ISP could have tens of thousands of users, all connected to their ISP, and all communicating with other networks too. At one point I managed to break into one extremely large ISP. I set up a sniffer there, and I continued the process, sniffing, hacking, sniff again, hack again, sniff again, hack again, until there were hundreds and hundreds of machines and their passwords in my sniffer logs and other hundreds I had already broken into as a result of the sniffer logs.

Here's what would happen to you if you checked your e-mail while I had a sniffer set up on your ISP's network: Maybe you subscribe to a local provider, one that serves mainly your area, like a local phone company. That provider has given you software, which you've installed on your computer and allows you to connect to the internet. When you dial up to your ISP, my sniffer picks up your password. If you use the internet a lot, maybe you even have accounts at other ISPs or other e-mail accounts at mail hosting faciles like Yahoo! or Hotmail. When you enter your password for these, the sniffer picks them up too. Maybe you checked your e-mail during the day when I was at school. Or maybe you checked it late at night when I was asleep, not that I was sleeping much. But that wouldn't matter either. My sniffer was functioning automatically, so when I got home from school, or got up in the morning, I'd check my log of captured passwords, and there you'd be. Then, using your passwords, I would break into your other ISP or mail hosts and do the same thing I did to the original ISP. And the process would continue on and continue on and continue on. I wasn't stopping for anything. I wasn't looking around, exploring systems, examining code. Nothing. I was on a March of Conquest. I wanted to have access to everything. I wanted to own the entire internet.

Sometimes I'd get caught. An administrator would notice a hack and find out that it came through a certain subscriber (you). He might fix the hole and close your account, or maybe just restrict your logins until he figured out what the problem was. But I was hacking in sessions. I would start when I got home, at three or four o'clock, and would continue on until next morning, knowing they most likely wouldn't see anything wrong until the next day. And even then they wouldn't know I had sniffed passwords for almost every user on their system. And you can't just change every user's passwords, the sheer number of them make it impossible. The only thing an administrator can do is restrict the account that initially got in, and hope the hacker doesn't come back. Besides, I could break into three, four, five hundred machines a session. Maybe fifty might realize they were broken into, and I'd lose access to those. But it didn't matter, because I still had three hundred machines that I could use for more hacking.

As I got more advanced I would have accounts at local ISPs that I had hacked. I would dial up using that account and from there I would connect to one of my safe machines. A safe machine is a computer I had hacked many months ago, one I've been able to maintain in such a way that no one has noticed I'm there, a machine I've got totally locked down. It might be an old university computer that's sitting on a rack somewhere, used by people to check e-mail but not really administered by anyone. It just sits there, unattended and unwatched. Except that I own it, and ten or eleven more like it.

So I get onto my safe machine, and from there I go to the machine I really want to hack into. If by chance an administrator for the machine I'm targeting discovers my hack, he's going to try to contact the person who owns the machine I'm coming from. But his attempt will fall on deaf ears, because no one cares about that machine, no one's responsible for it. So his effort will be fruitless—not like, for example, the University of Pennsylvania administrator's call to my old boss. And even if the administrator does get in touch with someone, there's no trace of my ever having touched that machine. I eliminated the traces by clearing my logs. In fact, I'd set it up so that when I logged in it never even generated a log entry. Like your alarm company setting up a special code in your alarm system so that it keeps no trace of they themselves ever having unlocked the alarm when they make a service call, for example. It's a back door.

Even if some extremely sharp security guy could trace my hack back to the safe machine, and even if he then traced that back to the ISP, the ISP would only go to some name I had made up for a fake account, or maybe to *your* name, because I had used your account to log on. And there would never be any connection back to me, exactly as if I broke into your house and used your phone to make a call. That call can be traced back to your house, but they have no idea I was the one who made it.

It was after many months of this kind of furious hacking that I got my second job, another one that just fell into my hands even though I wasn't looking. I mentioned to a friend of mine that I was in the market for a new ISP, and she said, "I'm using this one, it's pretty good." So as usual I decided to do my own research to see how secure it was. I broke into her ISP and a couple of days later they discovered the hack. *Well, they found out,* I thought. *That's a good sign.*

So I called them up and said, "My friend suggested you to me and I'm interested in signing up." And the guy's like, "Sure, I'll be happy to create an account for you."

"Oh, and by the way. I'm into UNIX. What OS are you guys running?"

"We're running FreeBSD."

"Really? Did you hear about that FreeBSD vulnerability that just came out?"

"No, I didn't. But we were hacked last week. Maybe that was how."

"Yeah, could have been." (I know you were hacked, I'm the one who did it.) "You should check it out. Let me point you to this web page where you can find out about the exploit."

"Cool. Thank you. You must know a lot about this."

"Well, I did security on my last job, at an ISP."

"Oh, really? Would you possibly be interested in a job?"

Between school and all the hacking, a job was the last thing I had in mind. But who knew? This might be a good opportunity. This was not the boss I was talking to, it was a technical person. But I knew he was thinking to himself, *This guy sounds like he knows what he's talking about. He's a technical guy, and it's difficult to find technical people, especially with ISP experience. Let's bring him down.*

Two days later I was at the ISP's office on Madison Avenue near Grand Central Station. This was the same kind of business as my previous job, with even more subscribers. But the offices were much smaller. Sasha's company had worked out of a huge loft (even if he lived in it). But this ISP had only three small rooms: one a closet, one the boss's office, and the third, slightly larger one for the administrators and business person. Half-finished computers and parts lay scattered all over the place. It was an amazing mess. I was sitting down with the boss, who was not technical at all, and he said, "Peter has never recommended anyone to me before, and usually he doesn't like the technical people we bring in. So the fact that he's recommended you means that you must know your stuff. He seems pretty impressed. I'd like to hire you; it's just a matter of how much you want."

I still had no idea what might be a decent rate, but I figured more or less double what I was making before would be good. So I said, "How about ten dollars an hour?" That didn't seem to be any problem at all. And it was legal, no kickbacks. So I was happy, though I also started thinking, *How much could I really make with this kind of work?* This was going onto the end of my second year at Performing Arts, the end of my third year in high school. And there I was, doing security and lots of other things (in an ISP you're a jack-of-all-trades), but still hacking away, wearing my white hat by day, my black hat by night.

This new job had fallen into my lap by accident, but now that I was working steadily and making money I began to think more seriously about the possibility of becoming a real security professional. Maybe hacking wasn't an end in itself. True, it was the most exciting thing I could imagine, but where was it going to lead, exactly? Besides, I had other things to think about. My mother was on my mind a lot, especially as I began to realize that she was suffering not just from her addiction, but from the disease. My mother was sick with AIDS.

For years now Mom had lived at home sporadically. Occasionally she'd be in jail for some minor drug-related offense, but more often she was out and living in one rehab facility or another. At the best of times she had her own little apartment and would try to find work. She was desperately trying to kick her addiction.

Through all of it, she had never abandoned us. Almost every day she could she would stop by Grandma's to see us and visit for a while. Over the years I had come to terms with her life and with my feelings. I knew she wasn't perfect. I knew she was struggling. But I also knew how much she loved us. All I could do was love her back and keep hoping that a change for the positive would eventually come, that she would get better. I knew she was doing everything she was capable of to straighten her life out, and as time passed it seemed to me that she was moving in that direction.

But at some point my brother Jevon and I began to suspect she had come down with AIDS. Her looks didn't say anything was wrong; she was still as beautiful as ever. But she started leaving pamphlets on AIDS around the house, as if she wanted us to know something, even if she couldn't say it directly. She also began hinting that she might not be around forever. By itself that might not have set us on the track, but together with the pamphlets the message sank in. No one spoke about it, not her, not Grandma, not Osie, but we knew. So it wasn't a surprise when she started going into the hospital for periods of time. Then, when I called once to find out which room she was in, Grandma was standing next to me. And when the receptionist at the other end had trouble finding her name on the patient list, Grandma said, "She might be in the AIDS unit." After that it was out in the open.

Once I really understood it, I suddenly felt a more urgent need to become successful. Money had always been on my mind. I had been playing around with entrepreneurial ideas ever since I learned how to work Mr. Forrest's Apple in junior high school. Part of that was my general idea that if I got rich I would be able to help my family. But when I dreamed about the actual things I would do with my money, it was always my mother I thought about. I'd have enough to send her to the best rehab program in the world with special facilities and the highest success rate. I'd have so much that I'd be able to move her and the rest of us out of the neighborhood and into our own place, far away from the troubles we were always dealing with. But now that I knew she had AIDS I became even more possessed by the idea of making myself a real career in security. Maybe that was where my money was going to come from.

■   ■   ■

One of the other employees at my new company had a boyfriend who was, or had been, a famous hacker. I didn't know him myself, but from what she told me he had done some fairly outrageous things. Now he was working for Gofish, the biggest new media consulting company in New York, one of the biggest in the country. I met him and I said, "Gofish sounds like a pretty cool place. You should hire me." And he said, "Yeah, maybe one day. Who knows." Basically, he blew me off. I saw him fairly often, whenever he came around to pick up his girlfriend, and it was always, "Could be. One of these days, maybe." Until it got to be almost a joke between us.

Then one night my technical friend Peter and I were playing around with a sniffer program he had written, looking at user names and passwords for accounts subscribers had at different web servers. We were checking out user names at porn sites, just for some after-hours amusement, and looking at various other things when I noticed a user connecting up to my co-worker's friend's company in clear rather than encrypted text. I was surprised that they would allow clear-text communication like that, so I joined IRC and sent him a private message: "Hey, I captured this session; I don't know if you knew people were connecting to your internal website in clear text. Thought you might be interested."

He got back to me immediately: "Thanks, I really appreciate that. Why don't you come down for an interview?"

Gofish had an entire floor in a huge SoHo office building. The elevator opened up to a young, hip work environment, as far as you could imagine from the cluttered, jammed-up office I was working in. Instead of everyone having cubicles, sleek metal slabs divided the desks. Music was playing softly somewhere, and it didn't look like anybody was really working that hard. They all seemed so relaxed and breezy.

We went into the conference room and the monitor there had a fireplace screen saver, which I'd never seen before, and I was thinking, *This has got to be the coolest place I've ever seen. This is what a high-tech company is supposed to look like. If this guy makes me an offer, there's no way I'm turning him down.*

We sit down at the table and he says, "Thanks for coming down here." I'm all ears. "And listen," he says, "I want to tell you, if you ever try to hack my server again, I have all your information." He's staring at me like a street thug who thinks he's intimidating a potential victim. "I have your social security number. I have your mother's maiden name. I know everything about you. You didn't think I was going to invite you down here without knowing everything about you, did you?"

I'm staring back, trying not to let the humor of this situation show in my eyes. I don't know for a fact if he had all of my information—he probably did—but I do know I had his. I had gotten it because I was curious about who he really was, with all of his girlfriend's talk about what a famous hacker he used to be. So we're locking eyes and I say to him, "You didn't think I was going to come down here without knowing everything about *you*, did you?" meaning that I had his identity info too, I wasn't just going to go down there blindly. He says, "Yeah?" We look at each other for a couple of seconds, waiting for the other guy to blink. Then he says, "Okay, come on, let me show you the office."

The reason our little standoff didn't come to anything is that both of us knew how easy it is to get somebody's personal info. All it took was a moment to establish that we both knew how, so what was the point of making a big deal of it? One way you could do it was to hack a credit verifying agency. Nowdays all you have to do is go online and pay nineteen dollars to get somebody's credit report, which has everything in it you might want, including social security number and mother's maiden name. But back then you had to get access by hacking in. Or you could get a lot just by going to one of the government's public databases.

If you wanted a different kind of challenge you could social engineer it, which put your acting skills to use. You might, for example, go to the person's long-distance carrier. Of course, to do this you had to know the operators' lingo.

"Hello, this is Judy from MCI, how may I help you?"

"Judy, hi. This is Tom from the New York office. Our TCAT

terminals are down and I have a caller on the line who wants to check his user account information. If you have a moment, could you possibly run a TC query for me? Caller wants to verify that his account information is correct."

"Sure, glad to help. What's the customer's name?"

The bottom line is that if someone really wants your identity information, he can get it. You can take precautions, but if somebody's intent on doing it, he will. Just like if someone wants to rob you or break into your house, he can do it, despite the fact that you might take care to lock your windows and doors. And once you understand that, you get to fear it less. There's hardly a point in stressing out too much over a fact of life.

■　■　■

I might be working in high tech and hacking frantically, but that didn't mean my entire life was online. I was never so completely immersed that I didn't have time for girlfriends, for instance, or for my other real-life friends, like Mark. I was also still living in Bed Stuy, which meant I was still getting into occasional situations. You just couldn't help it. One hassle was over a girl I was going out with. Aina was a beautiful Filipino girl I met at Performing Arts. This was in eleventh grade, and I had really fallen for her. We were in love, or at least I thought we were. Looking back, maybe it was just that I was in love. Anyway, when she broke up with me I was really crazy. Then I heard there was this new guy she started hanging with. This whole thing had my head spinning around in circles. Just the kind of thing to lead to a scrap.

"You know Aina?

"Yeah, I know Aina."

"You seen her?"

"Yeah, I saw her today, but she's gone already."

"Yeah, how do you know her?"

"What do you mean, how do I know her?"

We're squaring off already, and *whoom*, I get the first shot in, then drop down into a horse stance, waiting for him to come back at me.

He's saying, "What's going on? What's going on, man? What's up?" Taking off his jacket.

"Yeah, you know what's going on. You were playing with my girl." I'm just talking crap, trash talk. But my blood is racing. I can't believe Aina's really done this to me. I've just got to let it out. *Boom*, I hit him again and he goes down, but he's stronger than me and he grabs me and takes me down with him. He's punching me, I'm punching him. Two kids rolling around on the ground. Then he gets up. I'm still on my back, all the way up on my shoulder, trying to get him to stop pummeling on me. In the kung fu class I've been taking we've learned a little about ground fighting, so I'm trying to defend myself from my back. He's kneeling over me, throwing punches, when I see an opening and kick him right in the face, *bam*. And I see his face turn. One of the standard kicks, like a snap. *Pah*. And his face goes *whoo*. And by now a little crowd has gathered and I hear them going *oooohhhh*. He's dazed, staggering away from me. I go after him, but he grabs me again; I can see he doesn't want to fight anymore. One of his punches had broken my glasses and cut me. I'm bleeding. He's bleeding from the nose. Mark—he's there with me to get my back—Mark's yelling, "C'mon, man, let's go, let's go, let's go!" So we take off down to the train station.

That was it. It wasn't much of a fight, and I guess if one of us won it was me. But I didn't feel like my performance was anything to brag about. I knew my kung fu training had helped. While he had me on my back I was still thinking, I kept my composure, which is why I had seen the opening. Otherwise I would've just been reacting, rolling into the fetal position or just holding on for dear life. So it did help. But kung fu hadn't exactly turned me into a lethal weapon.

I'd started taking a real interest in martial arts my first year in high school, at Graphic Arts, after my introduction to aikido with the X-Men's fighting master. I was fourteen and I was in that really tough, scary school. Being there I'd sometimes get this feeling of helplessness I used to have a lot when I was younger, like a wave of weakness that would hit me in the stomach and spread. I'd think about how small I was, only 5'6" or so, and how skinny, and how I had hit adolescence early so I was pretty sure I wasn't going to grow much anymore. And meanwhile there were all these huge guys around, Decepticons and the others. I was always worried about what would happen when I got into a really bad situation, which was inevitable. There was going to be a day when someone came at me with a knife. Then what would I do?

I began to have daydreams about death. These death fantasies had actually started in junior high school, but they got stronger when I went to Graphic Arts. It was almost like an infatuation, something I just couldn't resist. I'd be sitting on the A train going to school and I'd start thinking, *What if this person sitting next to me pulled out a gun and shot me in the head?* I'd imagine how my blood would spurt out and splatter against the window, how the window would shatter ("They shot Blue in the head, his brains were on the sidewalk"—I had never really gotten that picture out of my head). I'd imagine how I would fall, what it would look like. All of this in slow motion. I'd see myself being pushed in front of a train, how it would happen, how the guy standing next to the pillar would suddenly turn and rush at me, bulling me over the edge, how I'd teeter there trying to catch my balance, how long it would take before I'd actually fall off the platform and land on the tracks. If I were shot and lying on the ground, how would I move, bleeding on the ground, how would I position my body in the final seconds, as the last breath left my lungs?

It wasn't the kind of paranoia that made me afraid to take the train. But it was there all the time, underneath. I'd think through my fantasy death scenes, like little movies in my head. What if this happened? What if that? I played out attack situations a thousand times. How would I react if some guy suddenly pulled a knife on me? Would I lean back, parry the knife, kick him in the stomach, get my back up against the wall? How *would* I react?

I'd think that if I learned how to fight, really fight, not just street fight, I'd have control. Then if I was in danger I'd know how to master the situation. I wouldn't be the victim, because I'd be able to defend myself.

With that in mind I bought some martial arts instruction videos, and I started learning from there. When I practiced to the videos the moves came naturally, almost as if I already knew them. In class—I was at Performing Arts by now—I was always showing people the moves that I learned. And one day someone said, "Hey, you know martial arts? My father teaches kung fu San Soo. Why don't you come down sometime and take a class?"

My classmate's father's San Soo school wasn't actually a dojo. He was just renting the space three times a week in the Alvin Ailey dance center. And he wasn't making a dime. The students were paying him enough to rent the space, but that was it. He was doing it for the love of teaching; he was dedicated to it.

In most of the traditional styles of kung fu you study forms for three, four, five years. Maybe in the fifth year you actually learn how to fight. But San Soo is different—it's all about fighting. San Soo is a truly devastating style of martial art. No forms, no set maneuvers, just fighting. Someone comes at you, you block it and use the leverage of the block to pull him into you. Smash your elbow into his Adam's apple, bring him down, knee him in the groin, push him back, break his knee, hold the arm, break the arm, then stomp his face. These are the basic moves of kung fu San Soo. There's none of the graceful, choreographed feeling of some of the traditional styles.

The day I showed up at the class, the first move I learned was: Someone throws a punch. Block the punch with the down windmill, strike to the throat. "In any situation," the teacher said, "if you can do that you'll stop an attacker. No matter how big he is." And it's a really simple move. Two steps. *Boom, boom.* I saw that and I was like, I've *got* to come here and learn this art. This is real street fighting.

San Soo is based on kung fu, but at least one of its most famous masters was a street thug. He killed people on the street using these moves. In China the real kung fu masters from the old days were all killers. There's no doubt about that. Even my current teacher will tell me, "Yes, my kung fu master in China killed people. That's why he had to run away to America, where he opened a kung fu school." The creator of kung fu San Soo had a lot of fights. He took bits and pieces of what he had learned from these fights to death on the street and created this style, what's called "free-hand fighting." You hit the body point to break it, you destroy your attacker. When he's on the ground you stomp him, make sure he's out. Then you get away. A very brutal art. Straightforward and brutal.

This was exactly what I was looking for. Every class involved sparring. In every class we'd learn a move. "Okay, the attacker's coming on to you, you kick at the knee in hopes of breaking the knee, come back down with a back fist to the temple in the hopes of knocking him out. If he doesn't fall, grab him and take him down. Okay, let's try that. Everyone grab a partner and let's try."

We did it in slow motion. Then we did it over and over until it got into our muscles. Then we did it some more. "Okay, someone's coming at you, first thing you do is move to the side and come back around like this."

I studied San Soo until I got good at it. It gave me confidence. But no matter how much you study, you are never actually sure if you know how to fight. You don't know if any of these techniques you're learning are really going to work in a street situation. You spar, but you're not really punching your opponent in the groin, you're not really breaking his knee or knocking him out with a back fist. You can't perfect blows like that in competition, you can't do them with gloves on. At the end of my street fight over Aina I came out more unscathed than the guy I was fighting, but I was winded and bleeding. I didn't feel like I had been able to do that much.

After that I began thinking that if I really wanted to learn to fight, the only way to do it was to do it. There's nothing that replaces actual experience. I had to know the feeling of the adrenaline rush, the feeling of exhaustion after the burst of energy gives out. And the only way you can learn how to recover from getting punished is to get punished. I knew I had to fight. If I was going to learn to defend myself, I had to get in the ring. So I began looking for schools where I could do that. What I really needed to do was go into competition.

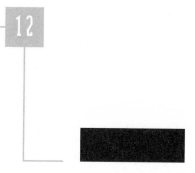

Meanwhile, I wasn't slacking off my furious hacking, what I thought of as my March of Conquest. Now I was breaking into government sites. Not that they were especially interesting, but hacking governments was the way you earned respect from your peers. Everyone was doing it at the time, or everyone with skill was doing it. So it just seemed a natural progression that I should do it too.

But being close to the government like that got me to thinking that maybe I would like to go into the military, or better yet, some spy agency like the CIA or NSA. I could become a hacker for the United States. Maybe even a hacker hero. What would that be like? I was going to have to do something after graduation, which was only a year and a half away. Why not defend my country doing what I did best? What an honor, if I got good enough and well known enough that the government would ask me to hack for it. On top-secret projects.

I had heard about some of those projects. Everyone deep enough into the scene knew that during the Gulf War a secretive group of hackers had been organized by "somebody" to hack Iraq. Iraq, including the Iraqi military, was still on an old networking system called X.25. X.25 was developed early in the networking era as a way for large numbers of different types of computers to connect with each other. It's still in wide use, but it's old enough that not many people understand it well anymore. Word had gone out in the underground that anyone who knew X.25 was invited to help. It was all shadowy, but the common knowledge was that the X.25 hackers had gotten into the Iraqi telco, the Iraqi banks, the whole infrastructure. There was even an article in one of the hacker magazines about Saddam Hussein having hidden money in banks, and the hackers having found it. The writer hadn't mentioned X.25 networks, but I knew that the hackers were these X.25 guys who were recruited by nobody knew exactly whom and organized nobody knew exactly how. But we can assume.

Anyway, I was looking around at different government agencies to see what they offered, what kind of jobs were available, how much they paid, that kind of thing. Sometimes I didn't even hack them, I just looked at their public web pages to see what information was available. One day I was looking at the ███████ site, reading over what it involved to be a ███████ officer, when I thought, *Why don't I just see what operating system they're using?* More out of habit than anything else.

I telnetted—a way of connecting up your own PC directly with another computer on the internet—into the ███████ machine.

And I noticed they were running a very insecure operating system, one that had default user names and passwords. By this time I was curious to look around, so I tried an account on this system that had no password set to it. I entered the user name—lp, for line printer (a standard default on that operating system), and next thing I knew I was in. It didn't take much more before I had root. I was amazed that ████████████ would be so completely insecure about its sites.

When I looked around the machine, though, I was disappointed. There was nothing interesting on it, just the public web site. So I disconnected. But by then I had plans in my head to change the web page sometime in the future. Lots of hackers do that kind of thing, and if they do it to a big website it's considered a coup. But I had never done it, and this would be an easy chance to get my name up; one of my names, anyway. This was going to be my one big public hack.

After that, every couple of days I'd spend some time designing my spoof web page. It had to be subtle and witty, something that at first glance would look like the real web page. But any visitor to the site who studied it a bit would realize with sudden delight that the site had been owned.

By this time I wasn't hanging with Digital Yakuza anymore. I'd had a falling out with one of the guys, nothing of any importance really, but I had more or less dropped connections with them. I wasn't a member of any group now, but I had started getting together with the 2600 guys. 2600 magazine is published by Eric Corley, and there's an associated hackers' radio show in New York, also called 2600. Some of the people around 2600 were famous hackers, or they'd probably prefer to be called "formerly" famous hackers, including Mark Abene.

Mark Abene's hacking career was legendary. As the notorious Phiber Optik, he had been a member back in the late eighties of the Legion of Doom, the most famous hacking crew of its time. But then Abene got kicked out of Legion of Doom; supposedly he had a fight with the top LOD guy, Erick Bloodaxe, aka Chris Goggans. After that Abene hooked up with some hackers in New York who had been exploring the telco systems. Abene himself was a real telephone expert, he might have known more than all the rest of them combined. He probably knew more about the telephone system than AT&T's own systems engineers. Phiber Optik and the others called themselves MOD instead of LOD. They were the Masters of Disaster, or maybe of Deception. They said it differently depending on what they wanted everybody else to think. They saw the telephone system as the hugest, most complex computer network of them all, and they hacked their way so far into it, just learning how the monstrous thing worked, that when half the country crashed on January 15, 1990, they were blamed. Actually, the Feds had been watching them for a long time, and had enough on them so that almost everyone in the group went to prison for a while. People are still going around wearing FREE PHIBER OPTIK T-shirts.

But it wasn't just Mark Abene; everyone in the 2600 group was well known and (depending on whom you asked) respected in the hacking community. Eric Corley ran the magazine and also hosted the talk show. He and the others had deep connections among hackers and hacking groups not only in the United States, but internationally. This was a great group to be part of and at some level I wanted to impress them, which was one of the things my upcoming ████████ hack was going to do.

I guess I was a little too eager about this, because one day at a dinner after the radio show I mentioned offhandedly that the ███████████ website was extremely vulnerable. I didn't say how or why I knew, and they didn't ask me, but everyone just took into account that it was, and shook their heads like, How could they be so lame? Except one guy, who said to the person sitting next to him, "Oh, I should look into that later." I heard him, but it was just a passing remark and I didn't make anything of it.

That was that and we went on to talk about other subjects. But two days later the ███████████ website was hacked, and the web page modified. And it wasn't modified with the web page I was creating, so it wasn't me. And the first site that mirrored it was the 2600 magazine site; that is, you could go to the magazine's website and view the hacked page. So everybody got a good laugh out of it.

But I wasn't laughing. It was my fault for telling them, and of course they didn't know what I was planning. But I still felt bad, like somebody had gone behind my back. No one ever mentioned anything to me either before or after the deed was done. But I knew it was someone who had been sitting there; I never told anybody else. So I was very upset about that. That was the last time I went out to dinner with them. No trust among hackers, I guess.

■　■　▦

It was right around that time that I was at work and noticed something on one of our machines that shouldn't have been there. Someone had logged on from a strange IP, an address I didn't recognize. Not only that, he was logged on as root. There were only five people who were root on that machine, me and four others. And the four others were sitting in the room with me, at their desks.

We'd been hacked, but it had just happened, because the hacker

179

was on the machine right then. He hadn't even had time to clean up after himself. His address was in the logs. It looked to me like he might even be coming from his home computer.

First thing, I disconnected him from our machine, closed down the account, and changed the root password. I knew every single thought that was running through this guy's head at that moment, because I'd been there. I even knew what he was saying out loud. He was saying, "Oh shit!" He saw that his connection was gone. Could possibly be a glitch of some sort, he hoped, even though he knew it wasn't. But still. He tried to log back on. But the account didn't work anymore. So now he knew for sure he'd been caught. It's just a matter of how serious his situation was. Maybe he was caught more or less accidentally by somebody who doesn't know much, who just lucked out and hasn't been able to trace him.

But, no, here comes a box popping up on his monitor that reads `root@machine.youhacked.com wants to talk. accept (y/n)` And now he's saying to himself, "Aw, man. They really have me. They got my IP address." Because he is by all means caught and busted.

I know he's afraid right now, that I really have control. He doesn't know if I'm going to call the Feds on him or what. So he accepts the talk connection, a form of instant messaging.

"Hi," I say.

"Hi."

"I'm the system administrator here."

"I'm the hacker."

"Yeah, I noticed. You just broke into my machine."

"Uh, yeah, I did."

He's thinking, *What's coming now? Threats? Some kind of*

flaming rant? Has he already notified the FBI? Then he reads, "How'd you do it? I thought I fixed everything."

"I used this new exploit."

"I thought I fixed that." (I've got the exploit in the directory.)

"Nah, you didn't."

"I didn't know FreeBSD was vulnerable to this."

"Yeah, it is."

"Good to know. Thanks."

"No problem."

"Well, uh, who do you know?"

"You're a hacker, too?

"I do associate with certain people."

"Certain people?"

"Well, yeah, I used to be in the scene."

We exchange handles; his is Pheno. I give him my old one, iGotz. But I don't recognize his and he doesn't know mine. As the conversation goes on, I'm maintaining an extremely elite aura. I caught him, so he's impressed. Maybe I know more than he does, maybe I'm better. Also, I have the power to turn him in to the police and lock him up if I want to. But it's pretty obvious I'm not going to do that, so I've got his respect for that, too, maybe even his gratitude. Of course he'd never say such a thing, but it's in his head.

Meanwhile, I know he wasn't breaking in to cause harm; he was breaking in because that's what he does. He isn't getting paid for it, he just needs to break into computers for whatever feeling of accomplishment he gets. So how is sending him to jail going to solve anything? It's not going to help him any, and it isn't going to solve the general problem of hackers hacking. It also doesn't benefit me in any way. I don't get a promotion for locking up a hacker. Plus I had already closed down the vulnerability. It's better that I find out how it was done and whether there was any damage, fix that, and go on with my life. One less hole. Other than that, I'm interested in what he knows, how his mind works. It's obvious that he's highly intelligent.

"Why don't you come into this channel on IRC?" he says, "channel" being the IRC equivalent of a chatroom. "There's a lot of cool people. You can meet Nekomon, my hacking buddy. He broke in here with me." Hmmm, so there were two of them, Pheno and Nekomon.

As I got to know them, talking on IRC, I could see that Pheno and Nekomon were really advanced. They had been hacking far longer than me and they had access to unpublished exploits I didn't have, partly because they had friends outside my circle. They also used techniques that were new to me, especially some really vast automated scanning and hacking tools they had written themselves. I got the sense that Nekomon had done a lot of the creative work and had taught Pheno, although Pheno himself was a first-rate hacker.

In one talk Pheno and I were having about unpublished exploits, I told him I'd give him an unpublished exploit if he'd give me one. Of course, I didn't exactly have an unpublished exploit right then, but I convinced him to give me his first. When he did, I put on a disdainful air and told him I already had it, how come he had given me something he should have assumed I knew about? Which saved me from having to give one to him. Not to mention sounding really elite, which I wasn't exactly.

On the other hand, I was getting better fast because I had been doing system administration for some time and I was understanding the technology in much greater depth. In my spare time I had also taken on the job of ports master for an operating system called OpenBSD. OpenBSD is a widely used UNIX-type system whose claim to fame is that it was created with security in mind. OpenBSD is a free system, which means that the source code is public (unlike Microsoft's, for example), so anyone who wants to write programs for it can. Developers attracted to BSD were taking programs written for other operating systems and rewriting them so that they would work on BSD systems—all this was free-form and voluntary, including my own work in managing the process of porting these programs from one system to the other.

So Pheno gave me this unpublished exploit, and he gave me access to a machine he had broken into and was now doing all his hacking from. He and Nekomon were showing me a lot of new things. They were giving me their custom code to cover my tracks, they were giving me sniffers for every type of operating system, they were giving me unpublished exploit code, which raises your hacking success rate. And now I had access to people I didn't know before, people who had even more code that I didn't know existed. And I'm being introduced to a new circle of friends, a much more talented group of hackers than I had known before. I'm now associated with Pheno, and Pheno has a reputation, so therefore I'm in. Pheno's a made man, and since I'm with him I'm becoming a made man.

Pheno didn't have just any friends, he was a member of w00w00. w00w00 wasn't a hacking crew, it was a hangout group from all over the country, but at the highest level. In w00w00 there were great C and C++ programmers, great assembly programmers, great UNIX engineers, great networking people. And almost all of them had at least some interest in security. From time to time w00w00 released security notifications to mailing lists like BugTraq, so in addition to everything else, w00w00 was a recognized security group.

After a while Pheno introduced me to these guys and I started hanging with them. But you can't just join w00w00. First you have to be recommended by a current member. When someone new is put up, the whole group starts to circulate questions. Do you know of this person? Do you like this person? What do you think of his reputation? Even if they don't know someone, people think, Well, Pheno's a good guy, we know Pheno, so let him come in. Or they might say, He seems like he's a good guy, but let's get to know him more before we decide.

The second requirement is that once you're recommended you go through a kind of probationary period for a week or two during which people are checking you out, asking you technical questions to verify that you know as much as you say you do. So you have to know what you are talking about. If someone points out that you're wrong about something and you're not willing to admit it, they'll call you a phony and kick you out.

I went through the questioning period without a problem. I talked about what I knew, and if I didn't know I didn't talk. But it's not just that you have to know what you're talking about, you also have to be willing and able to contribute information that will be new and interesting to the group. I joined in 1997. Membership in the group fluctuates and rotates, but the requirements are still the same today as they were then.

Most of the guys in w00w00 were college students, though some were still in high school and living at home. Others were out already, working as programmers and security people. We had no real agenda, no real purpose. We were just hangout guys. We still are. The best of friends, according to internet law. You know, your best friends from online are not equivalent to best friends in real life. You might never have met them, maybe you've never even talked on the phone, but you know how their brains work.

Now, five years later, everyone is working for major companies, major security firms, major financial houses. One is the head of security for Yahoo!; Napster's another old w00w00 guy. When I met Napster it was about a year, year and a half before he started creating Napster the company. He was just someone I was talking to online. No one special. Just another guy. Nowadays not many of the w00w00 guys have time to talk much unless it's a technical question that relates to their job. I was saying to one of them last week, "I miss the good old days when we could just hang out and talk about how much school sucked." And my friend said, "Yeah, work is just too crazy now." He's at one of the biggest networking companies in the world. We're just like what I imagine college grads are like when they get together to talk about the old school, or when they call each other for help with something. We're alums, too—of w00w00.

Pheno and I had planned to meet at DefCon, the big hackers' convention held annually in Las Vegas, although if they keep getting kicked out of hotels for rowdy behavior they might have to go somewhere else soon. Pheno was going to be there along with a bunch of other w00w00 guys, since DefCon also serves as the locale for w00w00's yearly get-togethers.

This was an exciting thing. Vegas—and on business, no less; I had gotten my company to pay for the trip. I'd been a member of w00w00 for a while now, and I knew most of the guys, either from digital photos or even from having met some of them in the city. They had told me what to expect at DefCon—thousands of people, lots of them script kiddies, lots of federal agents, lots of business types selling things, all sorts of hangers-on. But somehow I still had a particular vision in my head. DefCon was (and is) the biggest, most world-famous hacker meeting of all. The prime gathering place. And in the back of my mind I was seeing a kind of brotherhood of hackers, like some sort of immense throng of underground people emerging from their hidden lives once a year to come forth as their real selves. Almost mystical in a way. I went there expecting to see tons of talented hackers. I thought I'd be in the middle of hackers' paradise.

Instead, DefCon looked like a monthly 2600 meeting that had mutated into some kind of giant radioactive monster. Hordes of script kiddies and newbies and wannabees running all over the place, wearing black T-shirts and Doc Martens. Spiked hair, pierced noses and eyebrows. Just raucous. There were also lots of media people—the Aladdin Hotel convention hall was absolutely over-run by reporters and cameramen trying to close in on the action.

In its early years, DefCon might have been the annual gathering of computer mountain men coming out of their caves and lairs for three days of their own special kind of fun. But now it's been com-mercialized. I looked around, and with all the frenzy, the biggest thing I saw going on was business and promotion.

Booths were all over the place and everywhere you looked, peo-ple were being interviewed, everyone trying to get on camera. Thousands of people were doing whatever they could to get air time. People were selling how-to-hack books, companies were hawking all sorts of computer- and hacking-related products, security companies with intrusion detection systems were looking to test how their systems performed under constant hacking, and how many attacks they could find. Companies with computers and firewall products were there, saying, "Here, hack this machine" and everyone was trying to hack it so that afterward the company could say in its brochures, "We went to DefCon and nobody could hack us."

All around the hall people were making contacts, obviously selling themselves. This was not what I expected. This was not exactly a gathering of the sacred underground hacker brotherhood. At first I was totally shocked and appalled, though I noticed it didn't take me that long to get over it. I just started thinking, *I guess this is real life.*

DefCon is also famous for being infiltrated by federal agents, especially from the FBI and NSA. Which sounds stupid, because do they expect people to be hacking away at sensitive sites in the middle of the convention floor? (Well, okay, maybe it has happened.) More likely they're just there to see the faces and spot the latest trends in hacking, along with maybe seeing if someone on their most-wanted-hackers list is going to make an appearance. DefCon even has a famous Spot the Fed contest, in which hackers who think they've spotted a federal agent go up onstage and point him out. Then they call up the suspected guy and ask him onstage, "Sir, are you a federal agent?" And in most cases the guy says, "Well, yes, I am," embarrassed, or maybe getting into the spirit and having fun with it.

"Okay, let's see your badge."

So the guy shows his badge to prove it. Then the hacker who spotted him gets an I SPOTTED THE FED T-shirt, and the agent gets an I'M A FED T shirt. It is definitely one of the more popular events.

Other colorful events aren't hard to come by either. I was watching the hacking contest, in which the organizers set up a pseudonetwork and allow people to hack each other, when a woman climbed up on a table and started taking her clothes off to frenzied screaming and clapping from the onlookers. I don't think this was organized, many of the DefCon goers were like me, too young to get into a strip show, or go gambling either—not that that stopped anyone. But there's lots of free-form entertainment going on, some hacking-related, some not. There's also the scheduled stuff: talks, seminars, demonstrations on everything from cryptography to pirate radio to hacking tools, all of it along different tracks and set up for different levels. (The following year I gave a talk at DefCon myself, on how to secure distributed systems.) There's even a newbie-only track. But for most people, listening to talks isn't their primary reason for going. They go to see friends, make business contacts, and visit the strip joints and casinos. Associating with hackers might be the last thing on their minds.

I got used to the commercial aspect of it pretty quickly, but one thing that totally surprised me and changed my outlook on what it means to be a hacker was seeing hackers actively hanging out with federal agents, trying to get business from the government. Hackers who were supposed to be respected underground people, exchanging business cards with agents, looking to get consulting work—this wasn't at all what I imagined a hacker was; it was completely against the idea I had of it in my mind. These guys were doing business with Feds!

I was, of course, being very cool through all of this, even though it was a big deal to meet with w00w00—although in the end Pheno didn't show for some reason, so we never actually set eyes on each other, either then or later. One w00w00 guy, Jericho, ran attrition.org, the website that archives hacked web pages. He was older, maybe in his mid-thirties, definitely not young. He also hosted a popular mailing list called Internet Security News, which takes security-related news off of the Web and circulates it.

Later on, when I got to know Jericho better, I began to realize how deep his knowledge is about government agencies and their hacking groups. Jericho actually trained army, air force, and NSA guys on how to hack, even one of the army's red teams. One day I asked him about one of the groups he was teaching, "Are these guys any good?" He said, "Look, I'm teaching them what Linux is and how to compile exploits, so use your imagination." Jericho has this vast, strange hodgepodge of security information in his head. And he knows everyone. He may be the ultimately connected hacker. At the last DefCon I went to, people were walking around with Tsutomu Shimomura's password file printed on their T-shirts. Shimomura is the security expert who tracked down Kevin Mitnick (with a lot of help) in one of the most highly publicized hacking cases ever. I don't know for sure that Jericho had anything to do with printing Shimomura's password file on the shirts, but it's not impossible.

Jericho was only one of the w00w00 guys who were heavily into security. The last DefCon w00w00 meeting I attended was like a gathering of the crème de la crème of the security industry. But even at that first DefCon I was beginning to move seriously toward security myself.

By that time I wasn't going by iGotz anymore; instead of a handle, I had started using my real name. A hacker alias automatically associates you with doing illegal things. And I was already well on my way away from doing illegal things and into my work as a systems administrator and security specialist. So my handle by then was just Joey, short for Ejovi, and later evolved to what it is now—Joewee. I was on my way, I hoped, to establishing a strong identity as a figure in the security community.

What drew me to security rather than to going deeper into the underground was that I realized it was much more difficult to fix things than to break them. You don't really need to understand the technology to hack it. A great deal of hacking, especially now, is automated. As hacking tools become more sophisticated, they also become simpler to use. You don't have to be an expert to download a powerful hacking program and turn it loose, but you do need to understand the technology if you want to prevent hacking. So security has that extra level of challenge. Also, I felt that if I really wanted to be adept at programming, administration, and all the other work I was doing, I'd better have a good grounding in the underlying technology.

Besides, I thought that the best way for me to carve out a place in the security community would be to establish myself in my own identity. As I did that, people would respect me for who I am as opposed to my having the shadowy recognition of some code name. I wanted them to respect the real me, not the alias me.

So I was feeling strongly motivated. But at the same time, my motivation was being tested hard because Pheno was introducing me to all of these new hacking methods (Nekomon had kind of dropped out of the picture by this time; nobody said anything directly, but I got the feeling it was drugs, a problem for a lot of hackers). So I was suffering from temptations that maybe a saint could have resisted, except unfortunately I wasn't one.

Pheno was doing a lot of hacking, and at one point he hacked a machine he knew beyond a doubt was absolutely safe. It was like my old safe machines, but this one was locked down without any chance the hack would ever be found. So here was a machine he could hack from without fear. At the same time, I was experimenting with different exploits and I wrote one program that automated a particular kind of hack. That is, it would run through a list of internet addresses and hack each one of them blindly. To try it out Pheno gave me access to his ultrasafe machine and we set the script loose.

One day I was sitting there idly reading through the file of hacks the program had attempted. Most of them indicated "failed," since we had aimed it blindly at an extremely large number of IP addresses, basically the entire internet, and the hack would succeed only against the specific type of machine it was written to exploit. So I was kind of lounging back in my chair reading "failed," "failed," "failed," "failed," "failed." Most days I would have a couple of "succeeded"s thrown in, but that day there was only one. When I got into that particular machine and started looking around, though, I lost my casual attitude fast. I began flipping through directories and files, reading and reading. At first all I knew was that this was a ▮▮▮▮▮▮▮▮ machine belonging to some ▮▮▮▮▮▮▮▮ organization. But the hints started building up until I had an idea of where I was. And as I read more and more it wasn't just a word here and a word there, it was sentences and paragraphs. It was entire files. And I was thinking, *Hey, I know where I am!* And then, *Oh, man, we just got into* ▮▮▮▮▮▮▮. *What do we do now?*

13

The Holy Grail

A daemon is a program that attends to various computer functions automatically. If an e-mail is addressed improperly, it's returned by a "mailer daemon." The user is never aware of this mailer daemon. He or she wouldn't know it's there attending to mailing functions, except that once in a while it returns messages.

There are lots of daemons operating in computer networks, always in the background, like invisible spirits. One of them is a "status daemon," or "statd." The status daemon isn't meant for human viewing. It's a program that feeds information to other programs about what services are running and what ports they're running on, a kind of automated receptionist/doorman.

The status daemon runs on most UNIX operating systems, and the exploit I wrote targeted one of them, a popular Sun Microsystems operating system called Solaris. Solaris is used throughout the internet and has been for years, so it's a favorite with hackers.

The hack we sent out knocked on the door of every machine whose address was on our lists. It knocked, then it tried the key. If the door happened to belong to a Solaris machine, chances were pretty good the key would work, because this was an unpublished exploit, which meant nobody knew about it, so nobody had had a chance to figure out a way to counter it yet.

What the hacking program did was launch an attack against the status daemon. This hack was a buffer overflow that tricked the status daemon into ordering the computer to allow itself to be commanded from the outside; namely, from our safe machine. Sitting at my home computer, operating through our safe machine, I had full control, root, over every computer the exploit invaded, exactly as if I were sitting at that machine's terminal typing in commands.

■   ■   ■

I had a good idea where we were—a machine belonging to ▆▆▆▆▆▆▆▆. That was exciting, especially since one of ▆▆▆▆▆▆▆▆'s projects was all over the news. Of course, we were curious to find out about the project, but that could come later. First we just wanted to gain as much control as we could.

That wasn't hard. After I got into that one machine I set up a sniffer, and I used the same exploit from within the network to break into other machines. I ran commands that showed us the contents of the password files, all the encrypted passwords. Then I put the encrypted passwords into our cracker, and cracked them. Now I had fifty more accounts to use to log in to other machines. Then I did the same process on the other machines.

Meanwhile, Pheno was hacking in parallel with me. I'd be working one side of the network while he was hacking the other side. He was doing fifty machines there, I was doing fifty machines here. We'd go to those machines and from those we'd hack more. We cracked the passwords, set up back doors, sniffed other accounts to machines in other departments.

Every night I would check my logs to see what passwords I'd sniffed and what e-mails I'd captured. Eventually we controlled maybe 90 percent of the machines on this network. We had root access, we had every single password. On most machines we had three or four back doors to make extra sure we had access. We kept going deeper and deeper into ▮▮▮▮▮▮▮ until we reached the computers that actually controlled the ▮▮▮▮▮▮ that was all over the news. I felt like we were hot on the trail of the hackers's holy grail.

Most people would expect a high-tech organization like this to be more secure, especially one involved with such high-profile, sensitive projects. But that's not the case, not then and not now. These companies have a purpose; say, to develop a new ▮▮▮▮▮▮. To do such an infinitely complex task, the computer networks they use must perform at their peak. These systems are there to produce results, not to be secure, and the two goals often aren't consistent. Resources used to achieve security are resources deflected from accomplishing the mission. It's not that different from what goes on in businesses where the object is only to make a profit. If a company has a choice between hiring a salesman who might generate fifty million dollars in income or a security person who produces nothing, where it will put the resources seems like a no-brainer.

Also, these are very large networks. The task of securing all of these machines is almost impossible, especially since there is by design a lot of trust among the machines so that many people can be working on the same project, looking at the same data, and interacting easily with each other. You can secure the gateways—the standard ways of getting into the system—but if someone comes in through the back, you're just as insecure as you are from within. Recently there was a case involving the FBI in which an internal spy was using his access from within to go from computer to computer to steal information. Which proves that you might be safe from outside, but if someone can get in, he's in.

What we really had to worry about was the first machine we had broken into, the one we used to tunnel into the rest of the network. That machine had our scanning program in it, and every time we executed our exploit from our safe machine a log was created with our safe machine's address in it. So we had to make sure to remove those logs. But once we got into any subsequent machine we'd set up a back door, if we could, that allowed us to come in unlogged. And from that machine we could go to other machines. Even if we didn't have a back door we didn't have to worry as much about being logged, because connections between machines on the network looked like normal activity. They were always connecting with each other in the ordinary course of all the scientific activity that was going on. But even then we cleaned out logs, in case someone wondered, Why is this machine connecting to that one at three in the morning? As root? But once we were able to set up the back doors, they didn't log our connections at all.

The first thing was to get control of all the ██████████ networks we could. Once we had done that we started trying to learn more about where we were. We knew generally that we were in the machines that controlled ██████████. Now we wanted to find out in detail, so we started copying the source code to the programs that controlled the ███████████ that was in the news so much.

If the source code had been in one of the programming languages I knew, like C or C++, it would have made things easy. But the code didn't look like anything I had seen before. I knew that a lot of ████████████ still require the old computer languages to communicate, so I thought maybe this was a language that was used twenty or thirty years ago. Or it might have been flat text files for some really advanced compiler. I felt like some kind of archaeologist looking at inscriptions in a language he couldn't read. It was also confusing to me because it was all math. At least I was able to understand that it *was* math, and also that it had to do with ████████████ instructions of different kinds, but more than that I couldn't decipher.

But I wasn't actually that interested, either. What did interest me were the daily logs the project managers, engineers, and scientists kept so that every day I was able to see how they were progressing. I was immersed in the e-mail exchanges about the things they were doing, what they had in mind for this project, their talk about future projects.

As the days passed I felt more and more comfortable in the environment. I almost felt as if I were becoming a scientist myself. I mean, there I was, a part of the project every day, as if the engineers and scientists had a secret colleague watching over their shoulders as they controlled the ████████████. This was one of the reasons I loved hacking, for the times when I could become someone else and explore an exciting profession without ever having to leave my room.

Meanwhile Pheno and I were talking every day about the logs. I had an idea that he understood more about what we were seeing than I did, though he obviously wasn't any kind of scientist either. The fact was, even though we had been online with each other so much, I had no idea *who* he actually was. We had never met and he had never sent me a picture, so I didn't even know what he looked like. I assumed he was older than me, and that he might have some kind of high-tech job. But he could have been a fifty-year-old postal worker, or a black street kid, or an Asian techie. For all I knew he could have been a she, though I didn't think so. But whatever he was, the programs we were seeing were too much for either of us to grasp, even though we were on the system for almost two months.

I wondered, sometimes together with Pheno and more often by myself, if I couldn't possibly manipulate the ████████ some-how, despite the fact that I didn't understand the programs. The logs and memos I was reading were mostly end-of-shift-type messages, one scientist or engineer telling others what he had done that day—the problems he had had with the program, or with the environment, or with the ████████. The more I read these, the more comfortable I was becoming with what was going on. I might not have known what the programs said, but I was reading so much context that I was starting to have an idea what they did. And, of course, Pheno and I had root, which meant we could do the same things the scientists and engineers could, and even more, because not all of them had root access themselves. After a while I began to get the feeling that if I really tried, I could get some control of the ████████ myself.

I began to flirt with that idea. After all, it wasn't necessary for me to actually understand one of these programs. First of all, the ████████ scientists were compiling them, which means they were taking the language the programs were written in (which I couldn't read) and translating it into language the machines could read. So that was one big step that was already done for me. When I tried, I found I was able to run a program and export the display back to my machine—it would thus show up on my monitor and not the monitor of the hacked machine that was running the program. I did this a couple of times, just to do it. I was also able to look at the interface they were using to ███████████. But it was so mathematical that I didn't know where to start. I was looking at it, I knew what it did. I even understood some of it, based on the logs and e-mails that I was monitoring. But I wasn't quite there.

Pheno and I were talking all the time, coming up with theories on how things worked and what controlled what. We really wanted to do something. We knew that if we did do something, our names would go straight into hacker legend. Of course, there was also a good chance that if we tried we'd end up doing something we never intended. We didn't know, which was why in the end we decided not to touch it. Of course, if we had had a year to study the thing, we might have made a different decision. By then we might have felt knowledgeable enough to manipulate the project without any unplanned consequences.

The weird thing was that just after we decided not to do anything, they began having trouble with the project. Even weirder, right at that point we started losing access to the network. First one machine went down, then we lost a second. We tried the back doors, but nothing worked. I started getting a powerful gut feeling, like when you know somebody's following you down a dark street. You don't see him, but you know he's there. I knew they were coming after us. Pheno and I started saying that we really had to be careful of what we did. We made a promise not to discuss any of this, with anyone. It was just too high-profile, too completely dangerous and risky.

After the first two machines, we began losing the rest—though not all of them. It was obvious that their security people were monitoring the network. Now every time I connected to a machine, the next time I tried it was gone. We knew we might have made a mistake somewhere, no matter how careful we had been. Or maybe someone else had hacked into them and that hack had brought attention to us. Whatever had happened, there wasn't any doubt at all now. They were watching us.

Then out of the blue I got a call from an extremely well-connected friend of mine who always had feelers out all over the Net. "You're not going to believe this," he said. "There was a huge hack at ████████. The entire network was owned. They're really looking for who did it. If you happen to know who it was, I'd tell him to be very careful, because they're close to finding him."

With that we figured it was time to pull out completely. We even stopped discussing it between ourselves, so that the thing would eventually just die between us. We'd make sure to keep our systems clean, encrypt everything we had. There'd be no more hacking for a time. In effect, we went into hiding. Then, suddenly, Pheno disappeared.

We were used to talking to each other on IRC through a private connection. Then one day he wasn't there. And he didn't come back. I didn't believe he had been arrested. I also didn't think he had really left the scene, even for a while. The scene was who he was. And it's easy to disappear; all you have to do is change your handle and open a new account. Of course, he would want to keep all his connections with the w00w00 guys, but he could have had somebody introduce a new person to the group—the new person, the new identity, of course, being himself. People do this.

So Pheno faded into the ether. It was a little disconcerting to think that he knew more about me than I knew about him. He knew where I worked, of course, because we had met when he hacked our computer. All he had to do was call my company and ask for the person who handles security. Things were so casual there they would have just said, "Yeah, hold on, I'll connect you with Ejovi." It would be no real problem to find out my first and last name, and once he knew those he could get anything else he wanted. So I had to assume he knew who I was. But I didn't really think anything bad would come of that. If they got to me, that would only put them a little closer to him. So even if he had some knowledge, he wasn't about to use it. We really had the same interest—to keep this as quiet as we could.

Anyway, Pheno was gone. I stayed out there, but I never talked about the incident. I never felt compelled to tell anyone, or show anyone the logs. I didn't want to draw attention to myself or the fact that ███████████ had been compromised. Other than the anxiety that someone might be breathing down my neck, I felt good, like I had accomplished something unusual. Okay, I thought, I've done it.

As time passed, I got no unexpected visits. No vans parked in front of my house. No cars pulled up while I was walking down the street. Eventually I lost the feeling of my hair standing up on the back of my neck. I became a lot more cautious about attracting attention to myself. And I decided I was going to keep a low profile from now on and stop actively hacking. Besides, I was at the beginning of a serious security career. This business was definitely going to be my last hurrah.

But the underground part of me made it impossible not to fantasize once in a while about the kind of recognition I could have had. If I had taken the logs and published them. If I had published the programs. Even better, if I had really been able to do something with the project itself. Who knew—if I had only paid more attention to high school math, maybe I could have ███████████. That would have been the ultimate graffiti. After something like that, who could ever even think of bragging about putting his handle on some website?

■　■　■

Much later, after Pheno and I stopped talking and he disappeared, the *New York Times* site was hacked. Within my circle everyone had some idea of who it was. Nobody was really sure; the secret was well kept, but rumors were going around. And the likely suspect was the hacker formerly known as Pheno, current identity anyone's guess. The thing seemed to have his fingerprints.

In those days you heard several stories about the way the *Times* was hacked. The most common was that whoever it was had used a remote exploit—a web exploit, quite possibly a Solaris exploit, to break into one of the paper's web servers. From there the perpetrator was able to sniff around and break into the main web server and modify the *Times* web page. The exploit could easily have been, so they said, a buffer overflow, one that attacked the status daemon, though of course no one knew about that for absolutely sure. Or if they did, they weren't saying.

14

San Shou

I left the ISP in the middle of my senior year, mainly to get my
attention back on schoolwork during the homestretch leading
up to graduation. Then, just before summer, I picked up a consulting job at the New Museum of Contemporary Art. Nonprofits
are known for having out-of-date technology, but being a museum
of *contemporary* art, my new employer was eager to upgrade. My
job was to set them up with e-mail, bring their operating system
up to speed, and insure their internal security. It was a small gig
that I knew wasn't going to last more than a few months. I started
a month or so before graduation and worked through the summer, all the while trying to figure what I'd do afterward.

I kept telling myself, Well, if I can't find a job after this I'll just
apply to college. My grades weren't that good, and my SATs
weren't either, so I knew I wouldn't have a lot of choices. If college was going to be in my future it was going to be community

college. But I could apply and get in to community college any-time. There was no rush on that, so I figured I'd wait and see what might happen.

As the museum job got toward the end I began sending my résumé out, with the idea that it would be better to look for another job while I was still working. My previous ISP jobs had happened by accident, so I wasn't sure how this might pan out. I knew I had good experience, but what would companies think of a just-graduated high school kid who had never so much as taken a single computer science course? As far as I knew, most of them were looking for college grads. But maybe headhunters would know how to get around that, so I began sending my résumé to some of them. And right away I started getting calls back.

Some of my older friends in the tech world knew about job searches and they recommended a company called Otec. Otec wasn't a hacker recruiting company per se, but a number of peo-ple in the scene had gotten jobs through them and the company had a reputation in the underground as a good recruiter. So when they responded, I went to them first.

On my way to my first appointment at Otec I had in mind that my age was going to be a problem. But when I sat down to talk, it was a real awakening. The headhunter with whom I interviewed, a guy named Tom O'Connor, said, "Oh, you're how old? Nine-teen? Well, that's going to be difficult. When are you planning to go to college?"

"Well," I said, "next year, I guess," thinking, *What does that have to do with my skill set? Why don't you look at my résumé first?*

"Umm, yes, I see. Well, that still might be a problem for a lot of companies." He didn't seem that interested in me, as if he had already decided that my chances weren't too good. "I'll send your résumé out to a couple of places, and we'll see what happens." He said this very unenthusiastically. But suddenly he perked up. "Actually, there is one company that might not care too much about your background. Thaumaturgix. It was started by three technology people who do all the hiring themselves. There's no human resource person. It's a real you-go-there-you-do-technology type of place. So you might like that. And they might actually like someone like you." By the time I left the headhunter's office I was beginning to think that this seat-of-the-pants technology company might be my only real chance.

Down at Thaumaturgix's midtown offices I was supposed to interview with all three owners, but when I got there one was out so there were just two to talk to. I sat down with the first one, who introduced himself as Moses, very informal. His desk, actually not really a desk but a picnic table, had a Sun SPARC computer sitting on it, a machine favored by UNIX coders and engineers. That was an excellent sign, no lightweight Windows machines here. I was thinking that I liked this place already. We talked about my experience with UNIX, with security, my experience as a developer, my past. And he was like, "That sounds amazing. You know a lot about security and I think it's great. I also like your attitude. But . . ." Something was troubling him. "How old are you again?"

"Uh, nineteen?" (I'd told him already.) "But I'm going to be twenty soon."

"And you said you were going to school when?"

Everyone seemed to be fixated on this school business. "I'm going next year," I lied. I was trying to think up reasonable excuses for why I wasn't in college. "But I need to get some money before I can do that."

"I see, I see."

Then I interviewed with the second guy, Yogen (who also had a Sun SPARC on his desk), and I got the same thing—"You're how old? So, tell me, when are you planning to go to school?"

Eventually I found myself in front of the first boss again. The recruiter had said, "Whatever he asks you, don't mention how much you want to get paid." And now the first thing out of his mouth was, "How much do you want to make?"

"Well, you'll have to talk to my recruiter about that."

I was making fifteen dollars an hour on my museum job, so I knew I wasn't going to be making a hundred thousand a year. But I wasn't expecting it when the headhunter came back and said they had offered thirty.

I said, "Thirty thousand? I could make that working full-time at McDonald's if I wanted to."

"Okay," he said, "I'll make a counteroffer." He did, and got me an extra ten. I knew my boss at the museum was making thirty-five or forty, so that sounded pretty good. Forty thousand, for my first job out of high school. That was actually great! I wasn't complaining now. And with stock options, too! For the first time in my life I would be making real money. For me and my family. I wondered what Grandma would say when she found out.

When I started working at Thaumaturgix (TGIX, for short) there were 10 or 11 employees. When I left two and a half years later, there were 150. TGIX did consulting of all kinds, mostly for new media and start-ups, and that was a time when high-tech businesses were sprouting like daisies, so the work just flowed in.

From the start it was unbelievably hectic. When I joined, everyone, including the founders, did everything. There was no hierar-
chy; you just did what was needed and ten things were always
needed, usually yesterday. At the end of each totally flat-out day
we all released the tension by playing Quake, a game where you
can shoot and kill people over the network. So at TGIX you actually got to shoot and kill your bosses. Where else could you do
that?

TGIX was a great place because I learned everything. I mas-
tered new systems, because I had to implement them. And the
client always needed it right then, not tomorrow. So I had to learn
it right then, and be good enough to do it, not look like someone
who was struggling to stay half a step ahead.

Most of the jobs were hands-on hit-and-run gigs, like the time a
well-known food dot com was having trouble with the company
that managed its computer systems. The principals were not tech-
nological and knew nothing about their own systems, but when
the friction between them and their computer people reached a
serious level they decided to fire them. Problem was, they didn't
know how. The computer people had the guts of the business in
their hands, and who could say what might happen if they got
angry? When we analyzed the situation we decided with the principals that the best way to handle it would be to hack the dot
com's systems and acquire all the passwords and programs, surreptitiously, so that the computer guys wouldn't know it. We did
it, and when we were finished the principals were able to fire the
computer guys with some confidence that they would still be in
control of their own systems afterward.

Maybe the most exciting job of my two and a half years at
TGIX was the Empire Media caper, when we shut down Evans,

the deranged chief technical officer, before he could destroy the company. That was the kind of thing I really liked, an extreme environment that required speed and split-second decisions or you were dead. Jobs where you had a goal you had to achieve fast, and could see the results of what you'd done right away.

But some of the other, longer-range jobs were also cool. One time we were contracted by one of the biggest banks in the country to revamp a large part of their security, which meant installing systems and doing extended monitoring while we tested everything out and oriented the bank's computer administrators to the new environment. I was at the bank for two months, long enough to almost feel like an employee.

Every morning I worked there I'd put on my black leather jacket, my black scarf, and my sunglasses and take the train to Wall Street, where the bank's back offices were in one of the city's tallest buildings. This was serious business; at any given moment the bank was handling billions of dollars in transactions. But that didn't mean you had to lose your sense of humor. In my imagination I was one of the men in black, like in the Will Smith movie, the guys you don't know anything about, the guys who don't want you to know they exist, who never draw attention to themselves (of course, I'm in black leather and shades while everyone else on the elevator is wearing suits).

It's eight A.M. I take a seat at my desk up on the top floor. In front of me are the three computers and I've got a laptop in my bookbag for emergencies, plus my cell phone and wireless pager at my waist. I have a twenty-one-inch monitor connected to the three computers, each of which serves a different purpose. All of the machines are powerful, the most expensive money can buy. I turn on my monitor and log in. All over the city at this hour, at Merrill

Lynch, at Salomon, at Citicorp, at Chase, guys like me are sitting down and logging in. A small fraternity of cyberwarriors guarding the nation's finances. For the next eight hours I'm responsible for computer security at one of the world's most important financial institutions (along with my TGIX colleagues on this job), and the equipment we've installed gives us the ability to watch everyone in the company. We watch everything.

The first thing I do is check my e-mail. This morning I have forty alerts, various kinds of hacking warnings. I'm not worried. I know that maybe thirty-five of them are false alarms. The job now is to sift through the e-mail to figure out what's a false alarm and what's not; what's trivial, what's a real attack.

For the most part the alerts are the same ones I've been seeing every morning. They're a standard part of security at a place like this. The e-mail comes from a monitoring system we've set up to watch the entire corporation for suspicious activity. The alerts read something like: SUSPICIOUS CONNECTION FROM SOURCE [ADDRESS] TO DESTINATION [ADDRESS], giving me the source IP address and the destination IP address.

The monitoring is done by the intrusion detection systems, or IDSs, we've installed. The bank operation is so vast that we installed hundreds of them, each one watching its assigned segment, and all of them reporting to me and the others on our security team. The IDSs are giant sniffers dedicated to watching every single piece of information that goes across the network. It's like a monster tap on a phone line. The IDSs tap every computer in the company, thousands and thousands of them. We're monitoring every single employee who works at a terminal. If you work for a large corporation and you have a computer, you should assume you are being watched, too.

An intrusion detection system has the ability to send out messages—my security alerts—but it can't accept them. According to the network there is no such machine in the environment. An IDS doesn't have an address. It doesn't have an interface. It doesn't have anything. It just sits there, the unwatched watcher.

Along with firewalls, which guard the gates against intruders, intrusion detection systems are the most common tools used to secure a networked environment. They show you who's attacking, and the method by which you are being attacked. If you don't have an IDS, the only time you'll know if you're being attacked is if something out of the ordinary grabs your attention. Someone can be trying to hack you for a month before you start to notice anything, but if you have an IDS, then the moment something suspicious happens it's brought to your attention.

The IDSs are set up to flag a list of things we do not think should be happening—certain kinds of connections between networks, e-mail attempts from our company to various other companies. If any of these things happen, the IDS notifies us so we can look into them. In addition, the IDS checks for some commonly known attack methods and we decide if we want to investigate them. But there are some attack methods that are so common you don't really care about them. Scanning, for example.

The first thing an attacker usually does is scan your machines to see which services might be vulnerable. Nowdays there are so many programs you can download, everyone is scanning. People who have very little idea what they are doing can scan the entire internet. Even at home, where I have cable modem, I get scanned twenty or thirty times a day. I can't watch for that. If you're looking at thirty scans a day, you're going to miss the real attacks. So at the bank we don't care about the scans; we know it's happening.

On top of that, any serious hacker is probably scanning from a hacked machine. So even if we did trace it back, we'd find it emanating from a hacked computer, maybe in some place like Korea, Sweden, or China, or some other country where the computer's owner won't be responsive to a call from someone at a large American company who thinks he's being hacked. You go to an ISP in Korea and tell them someone is hacking you from their address, and you'll be lucky to get an e-mail back.

So pursuing scanners is just a waste of time. I'd never think of doing this kind of thing now, but when I first started working security and was more of a cowboy, I would hack backward to find the culprit. I would hack into the machine that did the scanning and see that the hacker was only using that machine, that he'd come from another one, maybe at some college. Tracing backward, using the same exploit he was using, I'd often be able to get to his original IP address. Then I'd get his information from his ISP account and give him a phone call, let him know we don't enjoy hackers scanning our network. Most often it was an American kid who would almost shit his pants when he got a phone call from someone.

But now I'm not interested in pursuing scanners. I'm looking for break-ins. So if I see an attempted buffer overflow, for example, from one machine on the bank's system to another, I'm looking into that. On my monitor I see: POSSIBLE BUFFER OVERFLOW ATTACK. SOURCE ADDRESS 10.10.10.21. SOURCE PORT 31337. Hmmm, I think, 31337—that looks like "eleet" (3 = E, 1 = L, etc.), a common hacker port. Besides, why would someone on one of the bank's machines be trying to hack another internal machine? That's a bad sign. That's something I want to investi-

gate. Now look, here's another alert: ATTEMPTED MODIFICA-
TION TO PASSWORD FILE. Someone's trying to modify the password file on a server, something else that shouldn't be happening.
So I start off with phone calls, calling the owners of those
machines to find out why those connections happened. My security day is off and running.

■   ■   ■

Working for TGIX, I was finally making enough money to start
looking for my own apartment. The last year or two living at
home had gotten harder and harder. Nothing had really changed
there. Grandma's house was the same overcrowded, noisy place it
had always been. Lots of love and laughter, but also lots of loud
squabbling and household demands that I didn't have the patience
for anymore. I was still sharing the same room with Jevon, so I
had no privacy, no place for the quiet concentration I needed for
my computer work, no place to hang out with girlfriends, no place
to just be by myself and relax. I wanted to get away from the
chores and the endless family arguments over who was supposed
to do what. TGIX was fun most of the time, but the work was also
tense and stressful. I needed to get myself free from the stress in
the rest of my life, and sometimes Grandma's house made me feel
like I was going to pop.

It was time to leave, go off and be my own man, even though I
knew Grandma was going worry about me. So I went to a broker,
who found me an affordable place up on 148th Street, right on the
border of Harlem and Washington Heights. In many ways I had
felt like an adult for a long time already, but when I actually
moved in the feeling really hit home. I had a serious job. I had my

own apartment. And I also had, somewhat unexpectedly, what looked like the beginning of a martial arts fighting career.

■　■　■

In the time since I'd left my San Soo school, maybe a year back, my martial arts career had taken some surprising turns—to the point where after a day spent fighting hackers I'd get to the gym, tape up, put the gloves on, and get into the ring against someone solid. At first all I was looking to do was get some ring experience, to see what real fighting was all about. I didn't want to be a professional fighter, I didn't want to be a sport fighter, I just wanted enough experience so I could be confident about taking care of myself on the streets. But somehow, very unexpectedly, that idea had gotten turned into something else entirely.

After I left the San Soo school I knocked around for a while doing some traditional kung fu. Then I started looking for a gym that had a fighting team. What I found was a place on Twenty-fourth Street called New York Kung Fu and Kickboxing, run by Steve Ventura and Dave Ross. Steve owned the place and Dave was head coach. They were both former martial arts champions, and Steve, especially, had tremendous experience, including having done a lot of fighting and studying in China. Their gym ran all sorts of classes, from aerobics to kung fu weapons training. But their basic idea was, if you want to learn how to fight, the only way to do that is to fight. Want to fight? Bring in some boxing gloves and let's get it on.

Steve and Dave specialized in Lama kung fu, one of the traditional Chinese martial arts. But their fighting team competed in San Shou, Chinese kickboxing, a form originally developed by the Chinese military for close-quarter combat. San Shou uses kicks, strikes, sweeps, and takedowns, mostly derived from kung fu and other traditional techniques. Besides that, Steve and Dave also taught ground fighting.

When I started hearing about all of this, I thought, *This is exactly what I'm looking for: punching, kicking, throwing and ground fighting.* That's pretty much what most street fights are about. Half the fights I've ever been in ended up on the ground. If it's one-on-one, then if somebody gets hit really hard his first reaction is to grab the other person and take him down to the ground to stop the punches. Or someone who doesn't know how to fight with his fists is naturally going to want to wrestle. So I was concerned about getting all these aspects of fighting under my belt as experience. And this was the only gym I found that offered all of those things in one place. So I signed up.

I started off with the idea of getting a couple of fights, just to know I could do it. And training for fights that went three three-minute rounds seemed about perfect, because most street fights end in fifty seconds. It's not like television, where they beat each other up forever before the bad guy goes down. In street fights, either some guy gets a good shot in and knocks the other guy out and that's the entire fight, or one guy is so much bigger or better that he just pounds on the other guy with the other guy covering up for dear life until the first guy gets tired of beating him or he can get away. I think I've only been in one street fight that's lasted a long time, and that was because we were boxing. And even then it was for at most two minutes. So I figured if I could last three three-minute rounds in the ring, kicking and punching the whole time, a fifty-second street fight should be no problem at all.

But looking back, I think maybe I was fooling myself about my motivation. Getting ring experience so I could handle myself better on the street sounded reasonable. It was a good justification. But as soon as I got into the gym, I got caught up in a totally different mind-set.

Steve, who owned the gym, was a great fighter, small like me, but fast and smart, and a great strategist. Outwardly he was a normal, friendly guy, but inside he was a driven fighter who had the curiosity and talent to master many different fighting forms. The kind of guy who, when he was a young martial artist, would hear about some competition and say, "What? Where?" Then he'd grab his bag and off he'd go. All he wanted to know was who he was fighting and when was he fighting.

I admired that attitude. As soon as I got involved I began thinking, *That's the type of fighter I want to be—someone who just says, "There's a competition? I'm there!"* Then, when they started telling me more about San Shou, it turned out there was a whole formal structure, including a U.S. national team that competed against other countries. I found this out the first week I started training, and I said to Steve right off, "Oh, so you guys have a U.S. team? I want to make it." And he said, "Well, you have to do a couple of things before you do that. You just don't walk onto the U.S. team. You have to become a national champion first, just to qualify." So I said, "Okay, I'll become a national champion too. Then I'll try for the world title." Steve laughed; he must have wondered who this was who'd just showed up at his gym. This was the craziest thing he'd ever heard. "Okay, yeah, okay," he said. "Just train and we'll see what happens."

But meanwhile, I was really pumped. I was saying to myself, I want to become a national champion. I want to become . . . whatever. Then I want to make the U.S. team, and then I want to become a world champion. And oh yeah, by the way, I want to become a professional and then a world professional champion. All this might have been nuts to listen to, but it was really just excitement coupled with something I always did when I got into something new: I always set the highest goals for myself. I can't do anything unless I'm aiming to become the very best at what I'm doing. I can't just get into something with no goal or purpose. I knew I had talent as a martial artist, it had always come naturally to me. So why not make the national team? Why not be a champion?

The training was hard and fast. The gym had bags and a regulation ring. I was there with some teammates who had been fighting for many years. I'd work on the heavy bag and it would be punch, punch, right roundhouse kick, left roundhouse, hook, kick, jab. For the first time I could feel I was really learning to fight. I sparred in the ring with Dave or Steve or my teammates, some of whom were so big and strong they could kill you. Duck, kick, cover, duck, move, jab, duck, kick, left, right, left. I worked like crazy preparing for my first fight, totally brimming with confidence.

The fight was at the gym, an International Kickboxing Federation regional event. I was fighting at 132, but when my opponent didn't show up it was decided to put me in against a 140-pound fighter. This was someone I had seen fight before, a two-time national champion, bigger and huskier than me, obviously stronger. On the other hand, my coaches knew his style well, and they thought I'd match up. My big advantage was that while we knew how he fought, to him I was a mystery.

When I stepped into the ring, a feeling of utter calm came over me, as if I didn't have any nerves. It surprised me a little, though it probably shouldn't have. I'd been training hard against guys on my team who were much bigger and stronger, experienced fighters, and I knew I could take a punch and give some back. Besides, I thought, this is no street brawl where you could get stomped or maybe knifed. In the ring there are rules and a ref. If you need it, medical help is nearby. There are things in life to be afraid of, but I instinctively knew that this kind of fighting wasn't one of them.

The bell rang. We moved toward each other and the guy plucked a kick at me. The first thing I thought was, *Has this guy ever kicked a bag? A person?* As we moved around and got into it, it was obvious he had skills, he knew how to throw the typical straight punches. But he seemed hesitant, almost afraid. I boxed and kicked, following him around the ring while he covered up, trying to figure out how he was going to defend against me.

After the first round, I knew I could beat him. I had won the round by a lot. Then the bell rang and I started doing the same thing to him all over again. Pushkick to move him back, jab to set him up, rear leg roundhouse to his head. *Wham.* Kick, punch, duck, move, just like in my sparring sessions. I was winning again, until halfway through the round I suddenly began to get tired, really weary. I could feel myself beginning to waste out.

The thing is, until you have your first fight you never know how hard you really need to train. You don't know what a real fight is like, and you have no idea that the adrenaline effect from training and that from fighting are two totally different things. Even when you're training to the max, you don't feel anything like the rush that comes when you're in the ring. The adrenaline just pumps through you. Then after a while it gives out, which leaves you in a state of fatigue you never experience in the gym.

That's one reason ring experience means so much. Once you get some fights under your belt you know this, and you try to pace yourself so it won't happen. Which is what my opponent was doing. He figured this is my first fight, I'm probably going to wear out. When I started getting tired in round two, he knew it. And then he started coming on. By the time we got into round three, I was just moving around the ring trying not to get hit. At the end I was ready to collapse—not from his punches and throws, but from exhaustion.

The decision was close. I had knocked him around pretty good up through the first half of round two, before he started coming at me. And even then he hadn't destroyed me or anything. When they announced it, it turned out I had lost by one point on all three judges' cards.

Afterward I was wondering how Steve and Dave had put me up against someone so experienced in my first fight. On the other hand, I knew they wouldn't have done it if they didn't think I could give a good accounting of myself. And the fact was, I had done really well, even if I'd lost.

The result was that afterward what I mainly heard was how great I was and how much potential I had. Suddenly Steve didn't think it was so preposterous that I might have a chance to become a national champion and maybe even make the U.S. team. I started to believe it, too. I felt really good. I had gone up against a champ and given him a hard fight. So even if it wasn't my original goal, I kind of got swept up into the whole be-a-champion thing. The fighting definitely took on a life of its own. I started tuning up for the nationals, which were only five months away.

15

Japan Dreaming

The way I saw it, the whole thing was simple: I was going to go to the Asian Martial Arts National Championships in Baltimore and I was going to win a title. All I had to do was prepare properly.

First I had a couple of local warm-up fights that went well and boosted my confidence even more. By now Steve and Dave were looking at the nationals as if they were only a stepping stone on the path to my future career in fighting. My job was to go to Baltimore and win in order to get ready for the next event. I wanted to make the U.S. team, that was the goal. And one of the criteria to qualify was being nationally ranked, having a national title. So the idea now was to get one.

A week and a half before the fight my mother came back from visiting her sister in West Virginia. We were all happy to have her home, relieved that she was alive after a scary hospitalization down there. While she was away, her AIDS had flared up and almost killed her. It wasn't the first time this kind of thing had happened. She had had the disease for years, and as it progressed she was suffering these bad episodes more and more often. Everyone knew she probably didn't have that much time left, but you managed to go on by hoping that as bad as it is, she would stay with us a while. It was hard to think about anything other than that she would be with us tomorrow, then the day after that, then the day after that.

But a few days after she got back, before I even had time to visit her at home, I heard from Grandma that she had gone into Woodhull Hospital, that she was really sick again. That Sunday Osie picked me up from my place in Harlem and we went to see her— that terrible Sunday when we got involved with the thugs at the hospital. It was an awful morning, staying with her, looking at her, seeing in her eyes that she was already gone, already focusing in on that place where nobody can go with you, but still suffering, not even being allowed to go in peace. At least we knew that she knew us, she recognized us. As weak and out of it as she was, she held onto us, not wanting to let go. I couldn't keep from crying, especially when I thought of how beautiful she had been, how smart she was and how her life should have been filled with promise, instead of this.

Three days later, on Wednesday, she was gone. We set the funeral for Sunday and everyone started helping with the arrangements, including trying to get the prison authorities to let my uncle Eric out for the day so he could be with us. The death of anyone in the family is hard, but your mother's death is different from anything else, so much harder. I was coping, dealing with it, mostly numb. Outwardly I was functioning, inside I hardly felt alive.

But I was also dealing with something else too—the fact that I had set a goal for myself of becoming the national champion. I'd been training for months. And to not go? Wouldn't that be the equivalent of giving up? And how could I give up when my mother had just died? She didn't give birth to someone who gives up. She didn't give birth to someone who walks away from a challenge. So I thought I had no choice but to go out there and win that title. That was my obligation. To win that title for her, to win it for myself. Because that was what I was meant to do.

The Friday of nationals the team rented a van for the trip to Baltimore. The wake was on Sunday, the funeral on Monday, so I would have to be back by Saturday night. I'd fight Saturday morning, be on the bus right afterward, and get back to New York to prepare. In my mind I wanted to come back with the gold medal and place it in my mother's casket. And tell her I had won it for her.

■　　■　　■

The Asian Martial Arts National Championships is a huge competition. At the University if Baltimore field house where they were held that year, six or seven tracks went on at the same time. Most of the sections were not fighting matches, but "forms." These forms—different varieties of what is called Wushu—are basically Chinese gymnastics performed both with and without weapons. In Wushu you see every kind of ancient Chinese weapon imaginable, from staffs to swords to pikes to wicked-looking long-handled axes. The moves are very beautiful to look at, though not practical for fighting, and not meant to be.

In today's Chinese martial arts, participants either fight or do forms. I watched the Wushu forms participants, all of whom—men and women—were extremely slim and graceful. They looked so delicate, though that was deceiving. They were doing amazing things, flying leaps and flips, and acrobatic choreography with sticks, or with these vicious-looking weapons. It was as if their strength was in their ligaments rather than their muscles. The fighters were different, almost like a different breed when you looked at them next to the Wushu people. A high-level San Shou fighter is cut, with ripped muscles in his back, chest, and stomach. It's very difficult to train for both fighting and forms, and you can see it instantly when you look at the body types.

The San Shou competition was divided by weight class. I was fighting at 132, at which there were only three competitors that year. Because I had almost beaten a national title holder in the weight class above me, I was given a bye. So I went directly into the finals. That was great, but when I warmed up I didn't feel so good. I had this strange feeling of numbness, of fatigue, as if my strength had deserted me. I was there, throwing combinations, warming up, but I felt lethargic. My punches didn't seem to have any power.

San Shou fights are held on platforms, square mats raised about two feet off the ground. The raised mat is a reminder of the origins of martial arts competitions in China, which were fought on open platforms, often twelve feet high. The idea was that the fighter would experience the same emotions he might have in a real fight—where death was always possible. If you got knocked off of one of those platforms, death *was* a good possibility. Nowdays if you get knocked off, you don't die. You just lose points.

When I got onto the platform I still had that strange feeling of numbness, but I forced it to the back of my mind and tried to focus on what I was going to do. Again, that prefight sense of having no fear came over me—I had no anxiety, no emotion at all, really. It seemed so simple being here, completely natural, like breathing. I put my mouthpiece in and looked over at my opponent, a stringy Russian who had beaten a tough, hard-punching fighter in his first match.

Hundreds of people were gathered around, watching us. I was staring at my opponent, and suddenly the referee screamed, "Fight!" The Russian came toward me. I threw a hard side kick to his thigh, so I scored a point, but his momentum pushed me back. He came at me again. I threw another side kick, but he was like a bull. His charge pushed me back again, so close to the edge I couldn't keep myself from going off the platform. I couldn't believe it. We were fifteen seconds into the fight and I had just lost three points.

I jumped back onto the platform and slammed my hands onto the floor, I was so frustrated. Not only did I come here to become the champion, I *knew* that I was better than this guy. I knew I was going to win. But here I got pushed off in the first fifteen seconds, which put me at a huge disadvantage. If I got pushed off again I'd lose the round automatically.

I might have been cool a minute ago, but now I was angry. I said to myself, Okay, it's not going to happen again. Not only is it not going to happen again, but I'm going to teach this guy a lesson.

As soon as I was back on the mat he came at me, thinking he was going to push me off again and end it quickly. I ducked under his arms, grabbed one leg, and jammed my other arm under his crotch, lifting him up over my head and slamming him down onto his neck. And I heard the audience go *ooooohhhhh*. I was looking down at him, pounding my gloves together and yelling, "Get up!"

The referee put us together again. "Fight!" Again he charged at me. I picked him up by both his legs this time. *Boom*, I slammed him onto his back, and crashed down on top of him, smashing my shoulder into his stomach, knocking his wind out. I jumped up, screaming, "Get UP!" The ref put us together. "Fight!" He came at me a third time. I picked him up and slammed him down again.

This time the Russian got up slowly. He was tired, beat. I could see the weakness in him, how much he hurt. When you watch fights on television, often you see someone get hit really hard and afterward you see him rebounding and maybe he gives the other guy a nod, acknowledging it, maybe even a momentary smile. The thing is, as a fighter you feel those shots, but you've got to move on. You can't think about it. The guys who think about it are the guys who get whacked one time and you can see it on their faces, they're like, *Shit, what did I just get into?* That's when they freeze, and that's when the other guy starts realizing, *Oh I got him now*, and *boom boom boom boom* that's the end of the fight.

That's exactly where I was now. I saw in his eyes that he'd had it, I could finish him. But all of this picking up and slamming was killing me. I wasn't pacing myself. My wind was totally gone, I was sucking air. I knew this guy was ready to go. His face was telling me, Now's the time. But I didn't think I had the strength to do it. And while I was standing there trying to dig it up, the bell rang and we went to our separate sides of the mat.

I said to David, my coach, "I can't feel my arms, I can't feel my legs. I feel weak." "Look, Ejovi," he said, "you got to get back in there, you gotta do this. You can't give up. Don't give up. You understand me? You didn't come here to give up. You go in there and beat him. You're killing him. Now suck it up! Suck it up and finish it!"

The bell rang and we got back into the ring. I felt as if my muscles had turned to jelly. I was so tired. You'd think that losing your mother would drive you to want the championship even more. I was hoping it would have that effect on me; I had been sure it would. But it wasn't. If anything, it was robbing me of my energy. The Russian came on. He wasn't feeling too good either, though he had obviously recovered a bit between rounds. We sparred, threw some kicks and punches, but my arms were lead, my legs could barely move. I was beginning to go numb all over. He came at me and threw a side kick. But it was slow and I grabbed his leg. With a huge effort I lifted him up and slammed him to the mat, going down myself on top of him. We scuffled around but he got over me and stood up first. I was so unbelievably tired. It felt as if I were paralyzed. I was thinking, *I can't do this. I really don't think I can finish.* But before I could make any decision, the ref counted me out. I didn't even know he was counting. I didn't hear him or see him. And by the time I realized what was going on, the fight was over.

■ ■ ■

It's common that fighters, once they get started, have something inside that doesn't allow them to quit. None of my teammates at the gym are professional fighters, but some have been in the ring for years. One, a lawyer, has been at it for a long time. He's good, a tough guy, strong and experienced, with a kick like a mule. But he's never won a title. On the outside that's what keeps him going, the hunt for a championship. But maybe the real story is that he hasn't proven to himself that he's good enough, or as good as he ought to be. So that's his internal struggle. Another teammate has had as many losses as wins, but his losses all happen the same way, at the tail end of his fights, when he gets knocked out or gives up too many points fast. What happens is that he loses control of himself for some reason, and suddenly he's not a fighter anymore, he's just a slugger. He stops thinking, he loses it. So he's really in search of that necessary self-control. That mastery of himself within. That's his need, his challenge. That keeps him fighting.

And for me, I've done things that I've later given up on. I wanted badly to be an actor, I wanted to be in the public eye. But it didn't work out and I didn't know how to keep pursuing it. At one point I wanted to be an elite coder, but I'm not. Hacker? Yes. Security specialist? Yes. Elite coder? No. I even wanted to be a rapper in my somewhat earlier days—though everyone in my neighborhood wanted to do that. But it didn't work out. I've wanted to be many things in my life and for various reasons I had to give them up. If I gave up fighting, I'd just be following the same road, the road of giving up in life. So I knew I was going to keep at it. Taking a silver in the nationals was a big disappointment, but it wasn't the end. I could still shoot for a national title, and then try for the U.S. team later. I didn't have to do everything at once.

■   ▓   ▓

As I started fighting again I also got to thinking more seriously about Japan. In America fighting and technology seem like a weird combination. Intellectual pursuits and physical violence usually don't go together. People don't expect it. I get that from co-workers all of the time. I mention to someone that I'm a fighter and most often it's, "How in the world could you do something like that? That's so barbaric. You're a technology guy, but you're a fighter too? That doesn't make much sense."

In Japan, it's different. I know in Japan people have much more respect for fighters. No one there gives a second thought to a person who's heavily engaged in martial arts, no matter what else he does. I began thinking seriously about the possibility of going to Japan to live for a while. In Japan they call San Shou *shootboxing,* and there's a huge amount of competition. They even have professional, not just amateur, levels, unlike in the United States, where the sport is just getting off the ground. But Japan attracts me in other ways too. The country has intrigued me ever since I was ports master for OpenBSD.

Actually, my interest in Japan started with a Japanese girlfriend, an artist named Kiomi whom I met in the East Village, where she was selling her paintings. That relationship didn't last long, but she started telling me things about Japan, and it made me curious to know more. So I began learning about the country, and it became an interest of mine. I didn't think Japanese technology was particularly exciting, but Japanese culture and language were. I started studying Japanese online. Then I met a couple of other people online who were also studying Japanese, who were also interested in the country, and we started talking and exploring things together. A little later I enrolled in a language school so I could get more serious. Eventually I planned to see Japan for myself.

My first trip to Japan was part vacation, part work. Before I graduated from high school, while I was still part-time at my second ISP, I was ports master for OpenBSD, a Unix-based operating system that is a favorite among many security-minded companies. In my spare time I was writing source code for OpenBSD and had taken on the voluntary job of "porting" programs others were writing; that is, taking programs that might have originally been written for other operating systems and making them available for any OpenBSD user.

Along the way I came across some Japanese software applications that I wanted to port, and I found myself talking (online) to a famous Japanese technology guru named Ito Jun, who was, among other things, an advocate of OpenBSD and a great expert at it. Ito Jun is such a fixture in Japan's high-tech world that he's known by everybody simply as itojun, small letters, no space.

Over time itojun and I became friendly, and after he got to know some of the things I was doing he introduced me to a Japanese group that was writing a book about OpenBSD. Not long after that, they asked if I wanted to collaborate with them. Apparently itojun had recommended me, and in the Japanese OpenBSD community, if itojun said somebody was good, that was the beginning and end of the discussion.

My section of the book was the introductory chapter, and it was while I was working on it that I took my first trip to Japan. By this time everybody at work knew I was carrying on a love affair with the country, and some of them probably didn't expect me to come back. I didn't know what to expect myself. All I knew for sure was that when I got to the Tokyo leg of my vacation I'd meet my writing partners and spend time talking with them about our project.

The last thing I expected was the reception I got when I stepped off the train in Tokyo. There, waiting for me on the platform, was a big group of BSD people. That night they threw a party for me, and everybody brought gifts. It was amazing. In the United States I was an unknown techie doing security for a start-up consulting firm. But in Tokyo I was being treated like a visiting celebrity. I couldn't believe it. My picture was staring at me from the computer magazines.

One of the things that opened my eyes was how wildly enthusiastic all of my new friends were about OpenBSD. OpenBSD was not my life; being the ports master was just something cool to do. After that I went outside, picked up my date, and went skating or something. But these guys were out of control. They spent their lives working around UNIX operating systems, and they were crazy for OpenBSD. "Our favorite operating system is OpenBSD!" "We just love OpenBSD!" "We want to know the future of OpenBSD!" "We want to know how to meet OpenBSD people!" They wanted to know everything. They were OpenBSD fanatics. That was a total shocker; it's still a shocker to me now. That's why the Japanese are so successful in so many ways—they get totally involved in whatever they do. And what these guys did was OpenBSD.

That trip made me want to go back, not to visit, but to stay a while. At least until I can get the language partly down and have some understanding of the culture. I'd like to hitchhike across Japan and see the most isolated areas. I want to know about the country's art and history. I'd definitely like to fight there, to see if I can make the progress I think I can.

After my mother died, this desire I had to go to Japan became even stronger. Her death magnified this sense I've had from my childhood that your time can run out before you know it, certainly before you're ready, that if you want to do something, you'd better do it now. See enough ODs when you're a kid, see people getting shot, and it makes you understand that "life is too short" isn't just a saying. The result is that I have almost no patience; it's one of my major flaws. Everything is urgent. I feel like every idea I have, every plan, has to be implemented as soon as possible. Otherwise I might never do it. In a way, the enthusiasm I attack everything with is my own private race against death.

I wonder sometimes if my mother would understand my desire to go to Japan. I think she would. As I am writing this she has been gone for more than a year, but so often I feel that she is watching me. I imagine that she can see what I have done, and what I am doing now, that she knows my thoughts. Maybe she can even see into some of my dreams, like watching them on a screen at a movie theater. If she can, then she probably sees me in the Japanese ring. A two-man show with hundreds watching. All eyes on us, the sweat pouring down my forehead, down my neck, and onto my chest as I move forward slowly, patiently, waiting for the opening. Or maybe she sees me traveling across Japan. Hitchhiking, the only way to get to know the countryside. I want to see the farmers who grow the rice. I want to work with them for a while, try and understand what this rice farming is all about. The soul of Japan, they say. I know I can see it clearly enough in my own imagination. The farmers will notice me coming down the road, not believing their eyes, apprehensive at first. But then they will walk toward me, the children laughing at the sight. A gaijin traveler appearing out of the sun of a hot afternoon.

Or maybe when I go there I'll just code, code and study Japanese. I'll wake up at two A.M. in a cold sweat and open my laptop. In the early morning hours my neighbor will hear the dull clickity-clack of my keyboard. At dawn the rising sun will warm my room with an orange glow that you can only find in Japan, and outside my window the crows of Tokyo will welcome me to another morning.

All of this might sound exotic, the dreams I am going to chase in Japan. But in some ways it isn't that different from what I've been doing for a long, long time now. I feel like I've been hacking foreign cultures most of my life.

16

September 11

Not long after my fight in Baltimore at the nationals I left Thaumaturgix for a security position at one of Wall Street's largest brokerage firms. For most people, this would have been just another step up the ladder. For me it was more than that; it was a kind of passage. For the first time I was really in the corporate world as a security professional. After all the underground stuff, the ISPs, even the cowboy world of TGIX, I had arrived someplace. Maybe not where I was going to stay—I was definitely going to Japan at some point—but this was surely different from the places I had been, about as far from Pulaski Street as you could get. Just being in this culture, where I am now, has been a hack. Wearing a white hat is something I had to figure out how to do.

In general you might say that inside everyone there's some anti-establishment feeling, at least at some point in their lives. Hackers all have a strong dose of this. By nature they are people who don't want to conform to standards, to what the world feels they ought to. So they hack. They hack technologies that would normally be used in one way, and they change them to be used in another. They think outside the box. If you're a good hacker, or even if you're just beginning, part of what drives you is that you don't want to do things the usual way, the way you're supposed to do them. Doing what you're supposed to be doing is not hacking; hacking is exactly about not conforming to the standards. It's about being different. Breaking rules.

People with that mentality are often antigovernment, anti–big company, antimoney, antisuits. They reject the idea that who they are might be defined by someone else, that they might fit into a slot. Coming to work for a large firm is the exact opposite of what these individuals stand for.

Part of this mentality is the search for fame, which motivates so many hackers. In the underground it's perfectly normal to want attention for every one of your actions, the opposite of the corporate environment, where you're expected to be a team player. But the hacker individual doesn't want to be a team player. A person from the scene wants to draw attention to himself.

One underground guy my old company hired would say things like, "If this company could hire someone like me, imagine what could be going on behind our backs." I asked him, "What do you mean, someone like you?" This was an actual conversation.

"Well, you know, dangerous, like me."

"Do you want the company to consider you a danger?"

"Well, I am dangerous. Give me sixty seconds and I can transfer money from any bank you name into another account."

"What are you talking about?"

"You give me sixty seconds and I guarantee that I can transfer money from any bank into any other bank in the world. Trust me, it's been done."

Now, this is ridiculous. The reason he's saying it is because he wants people to think, *Ah, this guy is really good.* That's great in the underground; people brag all the time on the IRC because they want to impress their hacker friends. But you can't come into a corporate environment and say that you can steal. It's not something you do. The need to draw attention to himself and flaunt his dislike of fitting in eventually got this individual fired. The day it happened, he came in with a big new tattoo on his arm, BORN TO DIE, wearing a short-sleeve shirt, of course.

But it's also true that the bad-boy rebelliousness of hackers doesn't always last. The media stereotype of the dangerous hacker, the antisocial nerd who can break into a bank in sixty seconds or solve the code to the vault with the bearer bonds, is just that, a stereotype, popularized because it sells. The truth is that within the underground there are many different types and that a lot of high-level hackers, even former criminal hackers—crackers—outgrow their past inclinations and now work for, or own, successful security companies.

What is very difficult is to wear two hats—to be one guy in the day, another guy at night. Some people, though, can be gray hats. They play both sides of the fence. I consider myself to be a gray hat now.

The thing about being a gray hat is that you have to know when to wear the white hat, when to wear the gray. I know that when I come into the office I take off my gray hat and put on my white hat, and when I leave the office I take off my white hat and put on my gray hat. Being a gray hat, I keep my hands in both pots. I still maintain my ties with the underground people, at least with my close circle of friends there. I might even write some programs specifically for them, such as scanners that would allow them to scan their companies' networks (they all have jobs, too) for common vulnerabilities. But then the next morning I go into work and I do my corporate job.

The fact is, if you are going to be a good security specialist, you'd better know what's happening on the other side. You have to stay connected. Otherwise you can't know the trends, you don't have your ear out for the unpublished exploits and the new attack methods. You lose your edge. And in this world of instant connections and global markets, losing your edge can have devastating consequences.

■　■　■

I had been working at the brokerage for almost two years. My office was near the top of the company's giant office building in Jersey City, New Jersey, overlooking the New York harbor, the Statue of Liberty, and the Manhattan skyline. The kind of place that really gives you the idea you can see out over the world. By this time I had moved back to Grandma's house, to save money for my move to Japan. I was going to go to Japan soon, but I had other things in mind. One was to buy a house for Grandma. The neighborhood around Pulaski Street was beginning to gentrify, which was hard to believe, but true. A few days earlier an old friend had told me there were only three black families left on his block. So I was thinking a lot about Grandma's place. It's rent controlled and owned by the city. She's lived there forever. But what would happen to her if they tore the building down and sold off the land? No matter what, there had to be a house for Grandma. A dozen other things were spinning around in my head, too. Maybe after Japan I'd spend a couple of months in a Thai kickboxing camp, and document it on video. Or maybe I'd take a job with a friend in Indonesia who owns a high-tech company and had been after me. But who knew, maybe after Japan I'd be ready to start my own company. I'd been thinking a lot about that, and about what kind of new product I could bring to the security field.

With so much to think about I was late for work one morning. I should have been out of the house by 8:30, but it was already almost nine and I was still getting my things together. In the background, Grandma's television was on, as always, tuned to one of the local news channels. As I was about to walk out the door I heard something about an explosion in the World Trade Center. They didn't really know what had happened, only that there was an explosion of some sort. They were checking it out.

I had to get moving. But this sounded like something I might want to take pictures of, so I grabbed my video and my 35mm cameras and threw them in my bag, listening to the news as I walked out the door. Outside of my building I could see smoke rising into the sky from the direction of the World Trade Center. The towers weren't visible from Pulaski, at least not from street level, but there was no question the smoke was coming from that direction.

Ordinarily I get the G train at Bedford and Nostrand, then the A, which takes me to the Chambers Street stop directly beneath the World Trade Center. Then I go into the Center and take the PATH train that travels beneath the Hudson River to Jersey City. But now they weren't stopping at Chambers Street because of the explosion. So I got off at Broadway and Nassau, one stop away.

As I was walking toward the World Trade Center with my cameras out to document whatever happened, all I knew was that something blew up in the building. I didn't know if anybody had been hurt. Maybe a few people died, I thought, worst case. I thought I remembered that five or six had been killed there in the earlier explosion back when I was thirteen.

Then I saw it. I was standing on the street with thousands of other people now, looking up at the building on fire, smoke pouring out. Now I could see that both towers were burning. I took a few pictures, but mainly I just stared up in amazement. People were saying that two planes had crashed into the buildings, that it must have been terrorists. Next to me a woman was crying. Another woman was praying to God, blessing the people who were killed or hurt. Tears were running down her face, her hands up in the air as if she were in church.

I was still a couple of blocks away, so I figured I should walk closer to get a less obstructed view. Now more and more people were crowding the street, watching. I asked one woman what she had seen. She said she saw fifteen people jumping out of windows. At least fifteen.

It didn't seem real. It didn't seem possible that this could really be happening. I was looking up, hoping to see something falling, to prove to myself that it wasn't people jumping, as she had thought, that it was just debris coming down. But I still couldn't see clearly enough, so I walked closer until I was right behind the Marriott, Financial Center just across the street. I took a picture of a cop on the corner roping off the street so people couldn't get closer. I took a picture of another cop holding back the people on my side of the sidewalk, next to a police car.

Right across the street from me, EMTs were placing someone strapped down on a stretcher into the back of the ambulance. One of the EMTs got into the back and closed the door. A little group of other EMTs were standing around, as if they were waiting for something. On the corner in front of the Marriott, a woman detective was giving instructions to other police officers—an Asian woman, with black hair and a gray trench coat, her police badge around her neck. Standing next to a gray unmarked cop car.

I was watching her giving directions, trying to coordinate people, when I heard a rumble. A sound I will never forget. It sounded like a plane flying overhead, like a rumble of thunder. And then the street started shaking. I looked up and the south tower was collapsing, showers of glass cascading everywhere. And the cop who was standing in front keeping us back was yelling, "Run, run, run!" And he was running too. So was the cop on the other side, who a moment before had been roping off the street. I turned around and took off, looking back over my shoulder. Crowds of people were running in panic, civilians, police, everyone fleeing for dear life. I couldn't tell if the rumbling was the sound of the building falling or the stampede of thousands of people.

On the run I took out my video camera and flicked it on. I didn't stop to turn around, but I pointed it behind me and shot. Then I felt a deep trembling under the earth and I knew for sure this was the building coming completely down. When I looked back over my shoulder, the whole thing was just a hundred feet away from me, the debris, rubble, glass, dust, smoke, everything. A giant black cloud of the building particles—a hundred feet back and coming at me fast. Then I fell.

I picked up my bookbag and my cameras and started running again. In front of me a woman tripped and went down. I grabbed her by the arm and pulled her to her feet. Then I was running again. I looked back and now the debris cloud was twenty feet behind me, getting closer and closer. I knew I wasn't going to be able to outrun it. It was coming down on me; already it was catching the people just behind me.

I looked for a door I could run into, an office or a store, and I saw one just across the street. A clothing store, people standing there with the door open. I sprinted across and into the store, screaming, "Close the door, close the door!"

The instant they slammed the door shut, the cloud swept by and the entire street went black. I was trying to catch my breath, not believing what had happened. I wondered about the woman who had fallen. She was a big woman. Not fast enough to outrun the storm.

With the street still pitch dark another woman came in, completely white with dust, a black woman but completely white. Then a second woman appeared, also covered in white powder, crying hysterically. With everyone else in the store helping her, trying to comfort her, I knew I couldn't do anything. In shock, I began to take pictures. Snapping pictures gave me a feeling that I wasn't really there. That I was behind the scenes somewhere, and that this really was a movie.

Once I got myself together I decided to go back to see if I could help. By now the black cloud outside was lightening up, turning grayish. I grabbed a couple of stockings—it was a woman's clothing store—went into the back, wet them, and put them over my mouth. Then I walked out into the street.

I couldn't see the towers at all. Down there it was still completely black. The street was dead quiet, no one seemed to be around. Then I heard a strange, low whistling sound. I could hear sirens too, obviously stationary, not phasing in or out as moving emergency vehicle sirens do. I saw two or three police officers coated in dust, then a couple of other people walking toward me out of the smoke, one man nonchalantly eating his breakfast as he came. I assumed he was in shock. He had a breakfast sandwich in his hand, and he was munching it as if nothing had happened.

I walked aimlessly trying to figure out what I should do, hearing that whistling sound, like the wind, the sound of wind blowing through trees. An eerie sound that I couldn't place. But I assumed it was just another part of this weird situation that was going on. The whistling sound, the stationary sirens, the cloud, the rain of dust—as if it were another world. Now I was walking toward the buildings, looking for someone to help, something I could do. On one corner there was a guy screaming into the cloud, "Walk toward my voice, walk toward my voice!" And people started emerging from the blackness. A man with his shirt ripped off, his arm bleeding. A woman with almost no clothes on. A man in pants but no shirt, covered in soot, bleeding from somewhere. Stumbling out of the darkness.

As I watched, the small trickle became hundreds of people. It hadn't registered yet that so many must have died, that what I was seeing were the survivors, walking away from the towers. And the man was yelling, "Walk this way, walk this way!" I put down the camera and started shouting myself, "Walk this way, walk this way!" Shouting at the people groping their way out of the blackness.

And while I was shouting I heard, or felt, the rumble again. The same rumble, the same trembling. And the sound of thunder, continuous thunder. I knew the sound, but it somehow didn't register that it was the second building. Again for a moment I stood there looking. And as I looked I saw a second cloud bursting out of the first blackness.

I turned and started running. Buses were lined up on the street, picking up people to take them away from the first building. I saw one of them packing up people and driving off. A second bus was waiting there, its door open. I ran into it, but it was empty. The cloud was coming down the street, just like the first one had. I tried to pull the door shut. But it wouldn't close.

I jumped out of the bus and ran into an office building, and again the street went black. I couldn't breathe. I sat down by the elevator bank on the floor, saying over and over, "Oh my God, Oh my God." Not just once, but a second time, before the smoke had even cleared, it had happened again. I thought, *This is it.* Now I remembered hearing on somebody's radio that the Pentagon had been hit. So when the second building fell and the streets of the city went black, I thought I would never see my family again, that I wasn't going to make it out alive, that this was the end of the world.

There were other people in the lobby who had escaped, all of them, like me, in shock. Outside it was black as night. I thought I'd never see the sun again. In war movies you hear the gunfire, you see the buildings explode. But to be there, to actually be there and to know that this was not a movie. To hear that sound. That thunder. To taste it in the soot. To smell it in the ash. To see it in the darkness out the windows. So real I could wipe it off my skin and feel it in my lungs.

And I sat there thinking, *Is this how it is? Is this the end of the world?* The Bible predicted that the world would end in flames, and I was sure the city was burning. I tried to call home, but my cell phone wouldn't work. I wanted to let them know I was okay, and that I loved them. But the phone would not work.

I stood up, looking out the door, praying that the blackness would go away. Promising myself that if it cleared I was just going to run and not turn around. And eventually it did clear. But I didn't run. Something forced me to stay.

I was afraid. The people who did this had struck fear into my heart, mine and everyone else's. They had stolen something from us. Our sense of security. Our sense of freedom. In the middle of a bright morning they had robbed us. But to run away would only be encouraging them. So I stayed, to document it, determined to recall what had happened. I had the feeling that if I survived, it would never happen again. And I wanted people to remember. And I didn't want to forget.

So I stayed. And as the smoke began to clear a little, I started walking toward the debris. And still I heard the whistling sound, and the ambulances and the fire sirens. But the sirens were moving now, no longer stationary. I could see lights coming from different directions in the smoke. And I saw policemen and firemen. But no one knew what to do. Not the police. Not the firemen. Because every single human being in that area was in shock. No badge, no uniform could ever prepare you for what had just happened.

And so I walked closer. Two blocks farther and the smoke cleared a bit, and I could see a silhouette of what used to be the center of the World Trade Center. A metal facade stood there, thrusting into the air, two hundred feet tall, blackened and twisted.

I took a picture. Then I put my camera down and stared in amazement and wonder, not that the buildings had collapsed, not that they were no longer there, but that this was all that remained. A blackened metal frame.

I watched as the newly arriving firemen came and looked in the same direction. I watched as the FBI agents and the police came down into the area, and as the EMT teams came. And they all stopped. And they all said, "Oh my God."

I went closer, until I was one foot away from that remaining structure, that two-hundred-foot frame. No one stopped me. I watched the firemen walking in the rubble shouting "Hello! Hello!" to anyone who might have survived, listening for sounds, for calls for help.

My eyes wandered to the corner where a gray police car, the same car I had seen earlier, was crushed flat by a huge beam. On the curb a fireman was sitting with his face in his hands, crying. I looked at him, wondering why he was crying instead of doing his job. I didn't know yet. I couldn't imagine.

I began to walk away. Among the many things I saw, under some debris, under some rubble, was a fireman's helmet covered in dust. And I thought, *Could this have belonged to someone who died?* But it seemed more probable that it belonged to someone who was running for his life and lost his helmet. I picked it up. I dusted it off and placed it on a piece of scaffolding. And I walked away, heading toward the Brooklyn Bridge.

A week later, looking through the pictures I had taken, I began to cry. I was thinking about the Asian woman detective in the trench coat next to the gray car that was crushed. I wondered if she was alive. I thought about the EMT workers and the patient who was strapped into that ambulance moments before the first building fell. Were they alive or dead? I thought about how close I was and how much debris I had stepped on, stepped over, walked through. Getting to the building, walking around the building. Walking up to it. I thought of how many people had died and the possibility that I had walked over some of them unknowingly.

How many might I have walked over without comprehending? And it was only then that I realized what I had seen. That what I was standing in front of was more than a two-hundred-foot metal facade surrounded by debris, it was a mass grave. And that eerie whistling sound. I knew what that was, too. A radio commentator had explained it, told how each fireman carries an alarm that senses movement, how when movement stops for thirty seconds the alarm goes off so others can find him. The whistling was the sound of hundreds of fireman alarms. It was the sound of death.

■　■　■

The next day, Wednesday, September 12, I went in to work to see if they needed me. I didn't think they would. As a security specialist it didn't seem likely there was much I could do. But I was wrong. Our main office had been located at One World Trade Center, the first tower hit. When the building went, it took with it all of the company's equipment, all the resources. Almost everyone had gotten out, but the technology was gone. We had also lost other branch offices within the World Trade Center area. Some were damaged by falling debris, others were just shut down, inaccessible.

All of the employees who had lost their offices had to be crowded into our New Jersey building, which, first of all, meant doubling up on desks. Beyond that the company had one absolute priority—to set up the traders and bankers so they'd be ready to do business when the financial markets reopened. Money had to come into the firm. That was the lifeblood. Everyone's job was to get the business side up and running, whatever it took.

We were told that trading would start again on Thursday, the next day. Everyone knew the CEO was frantically trying to convince the SEC and the other government agencies that trading should be delayed further, but while he was doing that, we worked under the assumption that we had twenty-four hours.

The first order of business was to get the traders behind PCs. Several thousand machines had been bought and shipped in overnight, a gymnasium-size room was already filling up with them. But each PC had to be "built"; operating systems, applications, the whole computer environment a trader needs, had to be installed.

We set up a room to do this and the moment new computers were unwrapped, people trundled them in and we started building. Our room looked out across the harbor, and as I worked, that day and the next and the next, I would look up and see the silhouette and the shadow of what used to be the Twin Towers, and the smoke rising from the ruins.

But meanwhile we built PCs. We set up a system where we first created an "image" machine, a primary computer that could use specially designed software to duplicate itself. The image machine was exactly what we wanted our PCs to look like, perfectly set up for what the business people needed to do. Then we connected ten other machines to the image machine and powered them up. As the image machine duplicated itself we started working on others so we could run additional networks. We were racing against the clock.

I wasn't a systems administrator for the company, but I was a tech person, so I could help with this; I could lend a hand to the systems people and the network people. Everyone in the room was working flat out, with occasional glances out the window at the destruction many of us had just survived. But you put that out of your mind, you compartmentalize it. What I cared about was only one thing—that the traders would be able to do their job when the markets opened. The fate of the entire firm was in our hands. Those machines needed to get built.

I worked until midnight, then went to my boss's nearby apartment and collapsed on his living room sofa. By that time we had probably set up a hundred machines. Another shift of administrators and engineers came on after us to keep the production going, and by eight the next morning when I started working again we had another hundred, enough for light trading, though not enough for full-scale business. But then came news of the reprieve. The market wasn't going to reopen until Monday.

With production gearing up I started looking at the security side, where we had lost servers and intrusion detection systems that guard the networks. Now the possibility of missing an attack was greater than ever. We worked fast to get the IDSs up to speed. We checked the firewalls. By now hundreds of employees were working from their homes or other outside locations, because there was just no space. So we had to set up security measures that allowed them to connect from anywhere, from home, from dial-up accounts, from internet cafés, from anywhere in the world, and still conduct business securely. To do that we already had devices that insured the safety of passwords, so that even if they were sniffed they couldn't be reused. I had built that system myself and we had deployed it with two thousand people. Now we needed to get it to five thousand more. We did that. We needed to set up redundant servers to authenticate passwords. We did that, too.

As I worked, I thought about what we were doing and about the new circumstances of our lives. None of the people around me had a job title anymore. No one was a security engineer, no one was a network engineer. Everyone was simply building the systems and networks that had to be built. Everyone was whatever he or she needed to be. All the formalities had dropped away; the only thing left was priorities. The room where we were building the machines was just down the hallway from the CEO's new office, and I'd see him walking out and coming back, struggling to save this company. This man whom I had never laid eyes on, whose name I would only see in print on occasion or maybe hear stories about at lunch. We never spoke, but it didn't matter. We were working next to each other, both of us, all of us, absolutely intent on survival.

It struck me that we were partners, teammates—words I never would have used before, that probably would have embarrassed me even to think. And I wondered if this would last. Does tragedy on this scale do something permanent to your feelings about others, about your dependence on them and theirs on you?

Personally, I felt needed, essential; emotions I have not experienced often in my life. In one way or another I have always been an outsider, coming in from places most people have never been to, through routes most people never take. I've been a hacker, someone trying to get in where others might assume he has no business being. But now it was dawning on me that my greatest hack had nothing to do with breaking into servers and networks. All that was petty, childish play. My greatest accomplishment wasn't finding a good technical job out of high school, something I had been so proud of. My greatest hack in life was simply to get to where I was now, to where I was part of something much larger than just myself and my own private dreams.

Late on Friday night I went back home, exhausted. Worried about the future, but also trying on these new feelings about who I was and what I had become part of, getting used to the idea that wherever it was I might be going from here, the truth was that in a lot of ways I had already arrived.

Glossary

Account

A name and password that identify an authorized user.

Address; IP address

The number or name indicating the internet location of a computer. IP stands for internet protocol, a set of rules that govern how computers contact each other.

Application

A software program that enables a computer to do a particular kind of job; for instance, a writing program, graphics program, or spreadsheet program.

Glossary

Back door
An unauthorized or secret way of gaining access to a computer or network.

Black hat
A hacker, a person who breaks into computers and computer networks. Can also mean a cracker, one who breaks into computers and networks with criminal intent.

Bot
Short for "robot." A computer program that operates automatically; for example, one that sends an advertising message to every user who signs on to a system.

Buffer overflow
A script designed to exploit vulnerabilities in software in order to execute unauthorized commands.

Carding
Buying merchandise with "phished," or stolen, credit card numbers.

Code
Instructions written for a computer in one of the various programming languages.

Compiler
A program that translates code written in a programming language into instructions a computer will be able to execute.

Courier

A person who distributes pirated software.

Credit card generator

A program that utilizes proprietary identification data from a credit card–issuing company in order to generate apparently valid credit card numbers.

Daemon

A program that handles low-level operating system tasks like keeping track of whether e-mail has been successfully delivered. Daemons run automatically and are not ordinarily seen by a computer user.

Default password

Password built into an operating system by its developer. Some software programs, for example, incorporate passwords that will automatically grant access. Anyone who knows such a password can access the system whether or not he or she is an authorized user.

Dial-up account

Each authorized user on a system is identified by an account, which usually consists of a name and password. A dial-up account is one that is accessed by dialing a phone number that connects the user to the system.

Directory

The UNIX term for a folder. A user can type in a command, called a directory listing, to retrieve a list of all the items contained in that directory. Directories can contains files and/or other directories.

Exploit

A program that takes advantage of a vulnerability in a protected computer in order to gain unauthorized access and/or control. A hacking program.

Firewall

A system of hardware and software that prevents unauthorized users from entering a computer or network.

File Transfer Protocol (FTP)

A protocol that facilitates the exchange of files between one computer and another over the internet.

Gateway

The authorized entrance to a network. A device that allows communications from one computer or network to enter another computer or network.

Gray hat

A person associated with both the hacking scene and the security world.

Hypertext markup language (HTML)

The language used to tell a computer how text and other elements (such as images) are to be displayed in a web browser. Its primary innovation was the ability to easily link one document to another. It is the core language for writing world wide web pages.

Image machine

A computer specially set up to replicate its operating system and applications on other computers to which it is linked.

Internet relay chat (IRC)

A specialized internet protocol that enables chat connections among users. More difficult to use than today's more common Instant Messenger system and much favored by hackers.

Intrusion detection system (IDS)

A computer that monitors all information going across a network, detects any unauthorized or irregular use, and alerts security specialists.

Internet Service Provider (ISP)

A company that provides subscribers with access to internet services such as the Web, e-mail, and news.

Log

A log is a record of particular events occurring on a computer system. One such log is called an access log, which is updated anytime a computer accesses a system. The record indicates such information as dates, times, and internet addresses of the connecting parties. Logs are often inaccessible by users without the highest levels of access, however, users with this access may modify or delete them (clear logs) at any time.

Newbie

One who is new to a field. In computer terms, includes those who, while mastering one aspect of computing, are new to other aspects—such as the internet or hacking.

Operating System (OS)

A program that manages all the basic functions of a computer.

Own

To have root access, or total control, of a computer or system.

PARC

Xerox Palo Alto Research Center.

Password cracker

Software that cracks or translates encrypted passwords into clear text.

Phishing

Scamming credit card numbers from unwary internet users.

Porting

Modifying a program written for one operating system so that it can run on another.

Promiscuous mode

When a sniffer is running on a computer, looking at all information that comes across the network to which that computer is linked rather than only at the information addressed to that particular computer, the computer is said to be set in "promiscuous mode."

Remote machine

A computer at a different location than a user's primary machine, but accessible through the internet or other means.

Root

Level of unrestricted access to all programs and functions on a particular computer. Having root is like having the highest level of security clearance. A user with root can see and control everything running on that computer.

Root password

The password that gives a user root access to a computer.

Safe machine

Hackers often do not hack from their own computers, but from computers they have previously hacked into. A computer of this sort that is not regularly checked by its owner (say, an e-mail server on a network) is called a "safe machine" because it is unlikely the hacker will be discovered on it.

Scanning

Rapidly surveying computers on a network, e.g., for vulnerabilities.

Script

Among the many types of languages used on the Web, scripting languages are quite common. A script, a program written in a scripting language, does not need to be compiled before it can run and is thus useful for tasks that require very little code to accomplish their task. Hackers frequently write in scripting languages.

Script kiddie

A low-level hacker who uses exploits others have written, most often downloading them from the internet.

Server

A computer that provides data or services to a network and users.

Sniffer

A program that monitors all traffic on a network and looks for specific types of information—passwords, for example. Like a telephone tap.

Source code

Instructions written in a programming language that must then be translated (or "compiled") into language the computer will understand ("machine language").

Status daemon

A daemon that provides information to programs running on a computer about which other programs are running and which ports they are using.

Telco

A telephone company

Telnet

A protocol that allows a computer user to sign on to and use a computer from a different location as if they were operating that computer physically.

Tool (hacking tools; scanning tools)

In hackers' terms, a small program that communicates with a remote computer in order to implement a specific kind of hacking procedure.

Trust

Links that allow users with accounts on one server to also use other servers so that information can be easily shared.

UNIX

An immensely powerful operating system, mainly used for network systems. Most of the computers that comprise the internet run this type of operating system.

Uniform resource locator (URL)

An address for a file located on the internet. Commonly associated with web addresses (e.g., http://www.website.com), although other internet services (FTP, for example) have URLs as well.

Warez

Hacker name for software, or "wares," traded or distributed on the internet without payment to or authorization from the producer. Pronounced *ware-ez*.

White hat

A person who counters hackers and implements computer security.

X.25

A set of rules, or protocol, developed early in the networking era that facilitates base-level communication between various types of public networks.